GUIDED WORKBOOK

LAURA IOSSI

Broward College

ELEMENTARY STATISTICS WITH INTEGRATED REVIEW

TWELFTH EDITION

Mario F. Triola

Dutchess Community College

Pearson

Boston Columbus Indianapolis New York San Francisco
Amsterdam Cape Town Dubai London Madrid Milan Munich Paris Montreal Toronto
Delhi Mexico City Sao Paulo Sydney Hong Kong Seoul Singapore Taipei Tokyo

Reproduced by Pearson from electronic files supplied by the author.

Publishing as Pearson, 501 Boylston Street, Boston, MA 02116.

ISBN-13: 978-0-13-449543-9
ISBN-10: 0-13-449543-8

1 16

www.pearsonhighered.com

Pearson

Contents

Integrated Review 1: Introduction to Integrated Review

Elementary Statistics Chapter 1: Introduction to Statistics

Objectives:

1. Learn how statistics is different.
2. Construct a plan of action for success.
3. Identify variables in context.
4. Recognize and convert between units of measurement.
5. Classify data as a type of number.
6. Evaluate exponents.
7. Translate numbers between scientific notation and standard form and vice versa.
8. Identify the number of significant digits.
9. Interpret technology output that is in scientific notation.

Welcome to statistics! In the corresponding chapter of the Triola text, you will begin the course with an introduction to the basic terminology used in statistics. Most students are not as familiar with statistics as they are with algebra. So, you may not be sure what to expect in this course. We will begin the *Guided Workbook for Elementary Statistics with Integrated Review* with a brief discussion of how statistics is different.

Since you have found your way to the *Guided Workbook with Integrated Review*, let me begin by explaining what this resource is and how you can use it.

This workbook parallels Triola, *Elementary Statistics,* 13th edition. The 10 integrated review chapters correlate with the first 10 chapters in the *Elementary Statistics* textbook. This workbook will review the algebra and arithmetic that you may need for the corresponding chapter in the text. For example, we will review how to evaluate a formula by substituting and simplifying in this workbook, and you will learn the statistical utility of the formula in the text. So, you may want to complete the integrated review for each chapter before beginning your work in the textbook. You may also wish to come back to this resource as you encounter difficulties with the algebraic and arithmetic calculations within the text.

The format of this resource is that of a workbook. The objectives that we will cover in each chapter will be clearly stated at the beginning of each unit. We will review each objective with an explanation or example(s). After each example, a My Turn! Problem will be provided. You will find space to work out these problems directly in the workbook. The answers to the My Turn! problems are located at the end of each chapter before the Practice Problems. The answers to all the Practice Problems are located at the end of the Workbook. In addition, you can find unlimited practice problems that correspond to this integrated review on MyStatLab. The Integrated Review problems on MyStatLab include corresponding help features such as videos, additional examples, and interactive Help Me Solve This experiences.

I hope that you will find this integrated review to be helpful, and I wish you success in this course. Let's start our journey!

Objective 1: Learn how statistics is different.

Most students join a statistics course not exactly sure what they have signed up for. It isn't algebra, is it? It definitely isn't calculus. So, what is it? I highly recommend that you open your textbook (either the paper version or etext) and peruse the chapter titles. You won't see terminology such as *factor* or *solve*. What you will see are such terms as *data*, *summarizing*, *graphing*, *estimating*, *inference*, and *probability*. Statistics involves making decisions based on information. You will learn how to gather, organize, and glean meaning from data. Yes, there will be calculations involved, and these calculations often involve large sets of data. Therefore, much of what you would encounter in statistics would best be handled with appropriate technology.

You may notice that many of the problems in the textbook appear to be "word problems." I find that this often makes students anxious. However, look closer at the problems. You are not going to be solving for x. The application problems that you typically encounter in the textbook are more interesting to read. The words are describing a real-world scenario. That is, they describe where the data came from and even provide the data we are dealing with. You may want to view the problems as a sort of information source from which you need to extract the desired information and numbers, rather like a dictionary or telephone book. (Do those exist anymore?)

Many students find that they enjoy statistics a lot more than a basic algebra class, as it feels more relevant to them. Statistics is everywhere. Once you delve into the course, you will be able to apply what you learn almost immediately.

Objective 2: Construct a plan of action for success.

As much as you may end up loving statistics, you probably want to have to take this course only once. As such, you should form a plan of action so that you succeed.

1. Determine your weaknesses and develop a strategy to overcome each of them.
   If you tend to lose focus or procrastinate, find a study buddy to motivate you.
   If you tend to leave materials for class at home, perhaps you can locate a locker on campus or have a prepacked statistics bag that contains your calculator, pencils, etc.
   If you tend to get anxious during test, find ways to simulate the testing environment as much as possible BEFORE you take a test.
2. Determine at least three ways to get help.
   What are you going to do when you get stuck on a topic? It is best to come up with several ways to get help before you are frazzled.
   Some ideas to consider:
   Watching the videos that accompany this text.
   Going to the math lab (It might be called something different at your college.).
   Visiting your professor during office hours.
3. Use appropriate technology.
   Statisticians use technology in the real world. If your professor allows it and recommends a particular technology tool, make the investment and gain access to it. Get your hands on

the technology as soon as you can. Practice with it. Use the same technology when you practice as you would during testing situations.

4. Determine an appropriate study schedule.
   It is generally recommended that you spend 2–3 hours studying/practicing outside of class for every hour that you spend in class. If you are taking this course online or in hybrid format, you will want to increase this recommended time. If you tend to struggle with math, you will also want to increase this time allotment.
   Also, remember to break the time into manageable pieces. The material will stick with you better in the long term if you look at it every day than if you have a marathon session the night before an exam.

If you need more study tips, there are entire books devoted to success strategies and you may want to consider finding one you like. For now, let's focus on our first objective of the integrated review!

Objective 3: Identify variables in context.

A variable represents an unknown quantity. In statistics, this could be a data value. You may want to consider using a variable name that is helpful in identifying what it represents (e.g., w for weight, h for height).

Note that within this course, you will feel as though you are reading Greek at times because you will be! A lot of Greek letters are used as variables in this course. The good news is that if you ever travel to Greece, you will be able to read some of the signs there. To see some of the Greek alphabet that you may encounter in this course, please see Appendix A.

Example 1 Let n represent the number of respondents to a survey about exercise. What does $n = 50$ mean?

It means that 50 people responded to the survey about exercise.

My Turn!

Let h represent the height of a wave in feet at "Jaws" in Maui. What does $h = 30$ mean?

Example 2 A six-sided die is rolled.

a) Let *d* represent the side that faces up. What are the possible values for *d*?
A six sided die has sides with 1, 2, 3, 4, 5, and 6 dots. So, *d* has the possible values of 1, 2, 3, 4, 5, or 6.

b) Let *t* represent the sides that face up that are multiples of 3. What are the possible values for *t*?
The faces of a die that are multiples of 3 are 3 and 6. Therefore, *t* has the possible values of 3 or 6.

c) Now, the die is rolled 4 times. Let *f* represent the number of times that the die lands with a 5 facing up. What are the possible values for *f*?
The die could land on a five 0, 1, 2, 3, or 4 times. So, *f* has the possible values of 0, 1, 2, 3, and 4.

My Turn!

Let *b* represent the number of boys in a family of 5. What are the possible values for *b*?

Objective 4: Recognize and convert between units of measurement

Since statistics deals with measurements, you should feel comfortable recognizing both metric and English units of measurement along with their abbreviations.

The metric system uses the following basic prefixes.

Prefix	Meaning
kilo	1000
hecto	100
deca	10
deci	0.1
centi	0.01
milli	0.001

The following are some common units of measurement in the metric system along with their abbreviations.

Metric	
Unit of Measurement	Abbreviation
gram	g
meter	m
liter	l

The following are some common units of measurement in the English system.

English	
Unit of Measurement	Abbreviation
foot	ft.
mile	mi.
ounce	oz.
inch	in.
pound	lb.

The above tables are not comprehensive lists, and if you are unfamiliar with a unit of measurement, you may want to look it up in one of our more comprehensive tables in Appendix B.

Occasionally, you may have to convert your data from one unit of measurement to another unit of measurement for the sake of consistency. That is, you generally want your data to be in the same unit of measurement before you perform any further calculations.

Here are a few of the most commonly used conversions.

12 inches = 1 foot
5280 feet = 1 mile
1 meter ≈ 3.28 feet
1 kilogram ≈ 2.2 pounds
1 mile ≈ 1.62 kilometers

Additional commonly used conversions can be found in Appendix C.

Example 3 Convert 72 inches to feet.

We will use the conversion factor that there are 12 inches in 1 foot. We write this as a fraction and multiply by it. We can do this, since the numerator is equal to the denominator; so, we are essentially multiplying by 1.

$$72 \text{ \sout{inches}} \times \frac{1 \text{ foot}}{12 \text{ \sout{inches}}} = 6 \text{ feet}$$

Answer 6 feet

My Turn!

Convert 5 miles to feet.

Example 4 Convert 8960 meters to kilometers.

There are 1000 meters in 1 kilometer.

$$8960 \cancel{\text{meters}} \times \frac{1 \text{ kilometer}}{1000 \cancel{\text{meters}}} = 8.960 \text{ kilometers}$$

Answer 8.96 kilometers

My Turn!

Convert 56 grams to milligrams.

Objective 5: Classify data as a type of number.

You will encounter several classifications for data in the text. Although we will not go over those here, we will indirectly practice classifying, using some concepts you probably know from prerequisite material.

Example 5 Select which of the following types of numbers the number 6 belongs to.

Whole numbers, integers, real numbers

Whole numbers = {0, 1, 2, 3, 4, …}

Integers = {…, −3, −2, −1, 0, 1, 2, 3,…}

Real numbers = all the numbers on a number line; this includes rational and irrational numbers.

The number 6 belongs to each of these sets of numbers.

My Turn!

Select which of the following types of numbers the number 2.5 belongs to.

Whole numbers, integers, real numbers

Objective 6: Evaluate exponents.

Recall that exponents are shorthand for repeated multiplication. The value of the exponent tells you how many times the factor repeats.

For instance, $3^4 = 3 \cdot 3 \cdot 3 \cdot 3$.

Example 6 Evaluate 8^3.

We expand 8^3, and then calculate the repeated multiplication. Most technology can calculate exponents, and you may want to explore how to do this on your calculator now.

$$8^3 = 8 \cdot 8 \cdot 8 = 512$$

Answer 512

My Turn!

Evaluate 7^4.

Objective 7: Translate numbers between scientific notation and standard form and vice versa.

Scientific notation is a version of shorthand for very large or very small numbers.

A number is in **scientific notation** when it is written in the form $a \times 10^b$, where a is a number greater than or equal to 1 and less than 10.

Example 7 Translate 0.000478 to scientific notation.

We have to move the decimal point until there is one digit to the left of it. (This is how you make a into a value greater than or equal to 1 and less than 10.) This would move the digit immediately to the right of 4. Count the number of places that we moved the decimal point and note the direction we moved it. We moved the decimal point 4 places to the right. Since we moved it to the right, the exponent on the 10 will be negative.

$$0.000478 = 4.78 \times 10^{-4}$$

Answer 4.78×10^{-4}

My Turn!

Write 0.0392 in scientific notation.

Example 8 Translate 760,000 to scientific notation.

For scientific notation, we have to move the decimal point until there is one digit to the left of it. This would move the digit immediately to the right of 7. Count the number of places that we moved the decimal. We moved the decimal 5 places to the left. Since we moved it to the left, the exponent on the 10 will be positive.

$$760,000 = 7.6 \times 10^5$$

Answer 7.6×10^5

My Turn!

Write 9,670,000 in scientific notation.

Example 9 Write 3.79×10^3 in standard form.

To translate from scientific notation to standard form, we are essentially doing the reverse of the above examples. Whenever the exponent is positive, we will move the decimal point to the right. Since the exponent in this example is positive 3, the decimal point will move right three places. (This is because we are multiplying by 1000.)

$3.79 \times 10^3 = 3790$

Answer 3790

My Turn!

Write 8.1×10^{-7} in standard form.

Objective 8: Identify the number of significant digits

The **significant digits** (or **significant figures**) of a number are the digits that carry meaning.

The following are considered significant digits:

1. All nonzero digits are significant
2. Zeros between nonzero digits are significant.
3. Leading zeros are never significant.
4. In a number with a decimal point, trailing zeros, those to the right of the last non-zero digit, are significant (they indicate that the number represents something that was measured to that level of precision).
5. In a number without a decimal point, trailing zeros may or may not be significant, and more information is needed to make that determination. (For purposes of our text, we will assume that they are not significant unless otherwise stated.)

Example 10 Identify the number of significant digits in the following numbers.

a) 938,000
9, 3, and 8 are each significant digits, since they are nonzero digits. The trailing zeros will be considered as insignificant for purposes of this text.
There are 3 significant figures.

b) 0.00012
The leading zeros are insignificant. The 1 and 2 are significant digits. So, there are 2 significant digits.

c) 0.000103
The leading zeros are insignificant. The 1 is significant. The 0 to the right of the 1 is significant, since zeros between nonzero digits are always significant. The 3 is significant, since it is a nonzero digit. So, there are 3 significant digits.

My Turn!

Identify the number of significant digits in the following numbers.

a) 1,030,000
b) 101
c) 0.000002

Objective 9: Interpret technology output that is in scientific notation.

Depending on the technology that you use, you may encounter scientific notation that looks slightly different on your screen than how we have been writing it thus far. You will want to try calculating something like 2^{50} using technology. The result is probably too large to write in standard form on your screen. On a TI-84 calculator, it may show up as 1.125899907E15. This output indicates that the number has been expressed in scientific notation. That is,1.125899907E15 is equivalent to $1.125899907 \times 10^{15}$.

Example 11 Round 2.058911321E14 (which is 3^{30}) to 3 significant digits.

2.058911321E14 is equivalent to $2.058911321 \times 10^{14}$.

If we round to 3 significant digits, we begin looking at the digits from left to right. The 2 is significant since it all nonzero digits are significant. The 0 is significant, since it is between two nonzero digits. The 5 is significant, since it is nonzero. This would be our 3 significant digit positions. However, for the 5 we have to decide whether to keep it or to round up. Since the digit immediately to the right of it is an 8, we round the 5 up to a 6.

So, our rounded value is 2.06×10^{14}.

Answer 2.06×10^{14}

My Turn!

The Excel output for 4^{30} is 1.15292E+18. Round this number to 3 significant digits.

Answers to My Turn!

1. A wave that is 30 feet high at "Jaws"
2. The possible values for b are 0, 1, 2, 3, 4, and 5.
3. 26,400 feet
4. 56,000 mg
5. Real numbers
6. 2401
7. 3.92×10^{-2}
8. 9.67×10^{6}
9. 0.00000081
10. a) 3 significant digits b) 3 significant digits c) 1 significant digit
11. 1.15×10^{18}

Practice Problems

1. Let d represent the days since a baby was born. If a baby was born on January 3 and today is January 15, what does d equal?
2. If t represents the time in years since the last U.S. presidential election, what are the possible values for t?
3. Convert 5 meters to feet.
4. Convert 35 kilometers to meters.
5. Select which of the following types of numbers the number -4 belongs to: whole numbers, integers, real numbers.
6. Evaluate 9^3.
7. Write 0.000706 in scientific notation.
8. Write 270,000 in scientific notation.
9. Write 9.71×10^{7} in standard form.
10. Identify the number of significant digits in the following numbers.

a) 503,000.2

b) 909

c) 3000

11. Excel output for $\frac{1}{4^{30}}$ is 8.67362E-19. Round this number to 3 significant digits.

Integrated Review 2: Decimals, Fractions, Percentages, and Graphs

Elementary Statistics Chapter 2: Exploring Data with Tables and Graphs

Objectives:

1. Round decimals.
2. Write a fraction in lowest terms.
3. Convert between decimals, fractions, and percentages.
4. Calculate relative frequencies.
5. Find the percentage of a number.
6. Plot points.
7. Review the skills for graphing.

In this chapter of the Triola text, you will be summarizing data using tables and graphs. In order to create these tables and graphs, you will have to be comfortable with dealing with numbers in several formats: decimals, percentages, and fractions. We will review how to convert from one number format to another in this *Guided Workbook with Integrated Review*. Also, throughout the Triola text, you will be expected to round answers according to various rounding rules. So, we will review the basics of rounding in this workbook. Finally, in the related chapter of the textbook you will learn about several graphs. To help you with this, we will review the basics of plotting points and the basic skills for graphing.

Objective 1: Round decimals.

Throughout the text, you will be asked to round your final answer. You may be asked to round in one of two ways: (1) to a particular place value or (2) to a certain number of significant figures.

Let's begin with a refresher on place value.

The table below indicates the place value for each digit in the number 1234.5678.

1	2	3	4	.	5	6	7	8
Thousands	Hundreds	Tens	Ones	Decimal Point	Tenths	Hundredths	Thousandths	Ten thousandths

You will often be asked to round your final answer to a particular place value. So, we will review the rounding basics in this first example.

Example 1 Round 3456.789 inches to the nearest tenth of an inch.

Step 1: Locate the digit in the desired place value. Let's call this digit R for purposes of our discussion.

The digit 7 is in the tenth place.

We are trying to decide whether the given number is closer to 3456.7 or 3456.8

Step 2: If the digit in the place value immediately to the right of R is a 0, 1, 2, 3, or 4, then keep R as it is and all digits to the right of R are zeros. If the zeros are not placeholders (that is, removing them does not result in other digits losing their place value), then they can be removed. If the digit in the place value immediately to the right of R is a 5, 6, 7, 8, or 9, then increase R by 1 (round up) and all digits to the right of R are zeros. Again, if these zeros are not placeholders, then they can be removed.

Since the digit immediately to the right of 7 is an 8, we must round up to 3,456.800. As the zeros are not placeholders, we can remove them.

Answer 3456.789 inches rounded to the nearest tenth of an inch is 3456.8 inches.

My Turn!

Round 7654.32198 miles to the nearest hundredth of a mile.

Example 2 Round 0.050748 cm to 3 significant digits.

We went over significant digits in the first chapter of this integrated review. Now, we will round numbers to a particular number of significant figures.

First, leading zeros are never significant digits. So, we will begin counting significant figures from left to right, beginning with the 5. The 0 immediately to the right of 5 is also a significant digit, since it is between two nonzero digits. The 7 to the right of the 0 is also significant, since all nonzero digits are always significant. So, by rounding to 3 significant digits, we are trying to decide whether 0.050748 is closer to 0.0507 or 0.0508. Since the digit immediately to the right of 7 is 4, we will keep the third significant digit at 7.

Answer 0.050748 rounded to three significant digits is 0.0507.

My Turn!

Round 0.36409 cm to 3 significant digits.

Objective 2: Write a fraction in lowest terms.

There are times that your result may be a fraction, especially when we work on the chapter on probability. Whenever you give a final result as a fraction, it must be fully reduced. In order for a fraction to be in lowest terms, there should be no common factors in the numerator and denominator.

Example 3 Write $\frac{48}{72}$ in lowest terms.

We may not all arrive at a reduced fraction by taking the same route. What is key is that we all check to make sure that there are no common factors left in the numerator and denominator. If there are any, cancel them out, since $\frac{a}{a} = 1$.

$$\frac{48}{92} = \frac{2 \cdot 24}{2 \cdot 46} = \frac{24}{46} = \frac{2 \cdot 12}{2 \cdot 23} = \frac{12}{23}$$

Note that many calculators (including the TI-83/84) can rewrite fractions in lowest form for you. You may want to become comfortable using this feature, especially for fractions with large numerators and denominators.

My Turn!

Write $\frac{36}{102}$ in lowest terms.

Objective 3: Convert between decimals, fractions, and percentages.

There will be many times in this course that you will want to convert between decimals, fractions, and percentages. We will review how to do this for the most commonly used conversions in statistics.

Example 4 Convert 8% to a decimal.

To convert from a percentage to a decimal, you will remove the percent symbol and move the decimal point two places to the left. (This is because the percent symbol means per hundred; so, you are essentially dividing by 100.)

Currently, the decimal point is to the right of 8. (An invisible decimal point is always to the right of the rightmost digit.) So, we will move it two places to the left. Since there is only one digit to the left, we will have to insert a placeholder of 0. Also, it is good practice to write a 0 to the left of the decimal point, as it results in less confusion to the reader. (Decimal points are pretty tiny for reading.)

$$8\% = 0.08$$

Answer 0.08

My Turn!

Convert 0.3% to a decimal.

Example 5 Convert 0.006 to a percent.

Now, we will do the reverse and convert a decimal to a percent. To do this you will move the decimal point two places to the right and insert the percent symbol.

$$0.006 = 0.6\%$$

Answer 0.6%

My Turn!

Convert 2.3 to a percent.

Example 6 Convert $\frac{7}{8}$ to a percent.

First, we will convert the fraction to a decimal. Then, we will follow the steps from Example 5 to change the decimal to a percentage.

To change a fraction to a decimal, you divide the numerator by the denominator.

$$\frac{7}{8} = 7 \div 8 = 0.875$$

Now, we will convert the decimal 0.875 to a percentage by moving the decimal point two places to the right and inserting the percent symbol.

$$0.875 = 87.5\%$$

Answer 87.5%

My Turn!

Convert $\frac{9}{16}$ to a percent.

Example 7 Convert 0.06 to a fraction.

To convert a decimal to a fraction, you write the number with the decimal point and leading zeros removed as the numerator of a fraction over a denominator that is a power of 10. The exponent on the 10 is the number of digits to the right of the decimal point in the given number. For 0.06, there are 2 digits to the right of the decimal point. So, the denominator will be 10^2, which can be rewritten as 100. Once you have a fraction, be sure to simplify it fully.

$$0.06 = \frac{6}{10^2} = \frac{6}{100} = \frac{3}{50}$$

Answer $\frac{3}{50}$

My Turn!

Convert 0.0028 to a fraction.

Objective 4: Calculate relative frequencies.

As you construct tables and graphs in the text, you will be calculating relative frequencies, which are the ratio of the amount of times an event occurs to the number of times the event could occur. In other words, it is the number of times something happens divided by the total number of times it could happen. Relative frequencies are often represented as percentages.

Example 8 If 30 out of 80 people responded that they skip flossing at least one day a week, calculate the percentage of respondents that skip flossing at least one day a week.

So, the relative frequency for this problem is 30 out of 80, which can be written as $\frac{30}{80}$. We can then turn this fraction into a decimal. Finally, we convert our answer to a percentage by moving the decimal point two places to the right and inserting %.

$$\frac{30}{80} = 30 \div 80 = 0.375 = 37.5\%$$

Answer 37.5% of respondents skip flossing at least one day a week.

My Turn!

If 450 out of 800 people responded that they own more than one computer, calculate the percentage of respondents who own more than one computer.

Objective 5: Find the percentage of a number.

Sometimes, you may need to determine what the percentage of a number is.

What is $p\%$ of x?

To do this, you will multiply the percentage written as a decimal times the number.

That is,

$$p\% \text{ as a decimal} \cdot x$$

Example 9 What is 30% of 150?

Write 30% as a decimal and multiply it by 150.

$$30\% \text{ of } 150 = 0.3 \cdot 150 = 45$$

Answer 30% of 150 is 45.

My Turn!

What is 65% of 600?

Example 10 There were 900 respondents to a survey. If 40% of respondents indicated that they live in an apartment, how many respondents live in an apartment?

This amounts to answering the question, "What is 40% of 900?"

To find the percentage of a number, you multiply the percent written as a decimal times the number.

$$x = 0.4 \cdot 900 = 360$$

Answer 360 respondents live in an apartment.

My Turn!

There were 80 respondents to a survey. If 30% of respondents indicated that they own a DVD player, how many respondents own a DVD player?

Objective 6: Plot points.

In statistics, we will often study data that come in pairs. It can be helpful to see the paired data on a graph. Each set of paired data can be represented by a point that can be written as (x, y). The points can be plotted on a Cartesian coordinate plane. The origin, or the place where the horizontal (x) and vertical (y) axes intersect, is the starting point The first value in the paired data (the x value) will tell you to move left (negative) or right (positive). The second value (the y value) will place the point up (positive) or down (negative). For example, you can plot the set of paired data (3, 5) by counting three units to the right of the origin and up five units.

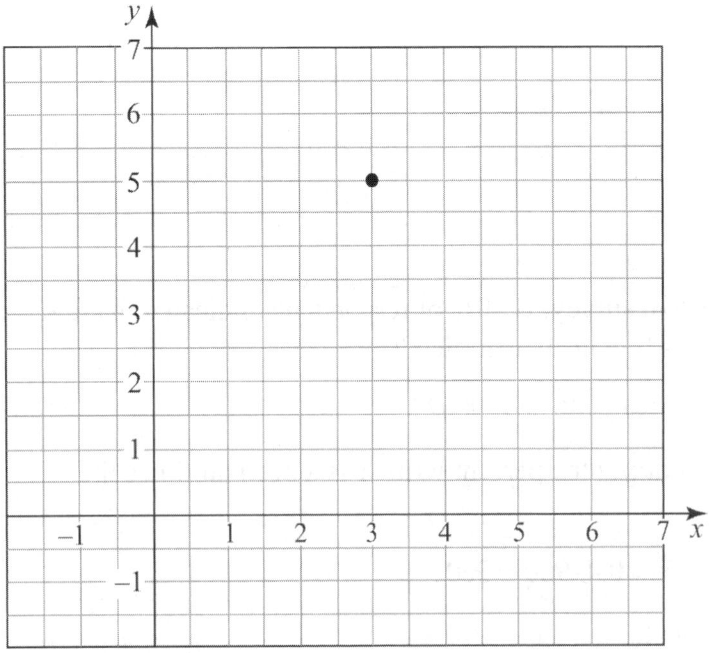

Example 11 Plot the following paired data.

Number of People in a Household (x)	Number of TVs in a Household (y)
1	2
2	4
3	5

These data can be written as three points (1, 2), (2, 4), and (3, 5). Then, we can plot these three points on the Cartesian plane. As you learn statistics, you will find that most of the data that you will encounter will be positive values. As a result, many times you can limit your Cartesian plane to the first quadrant (the upper right-hand section). Also, there may be times that you will want to change the scale on the axes. Be sure to label your axes with the variables that are being used.

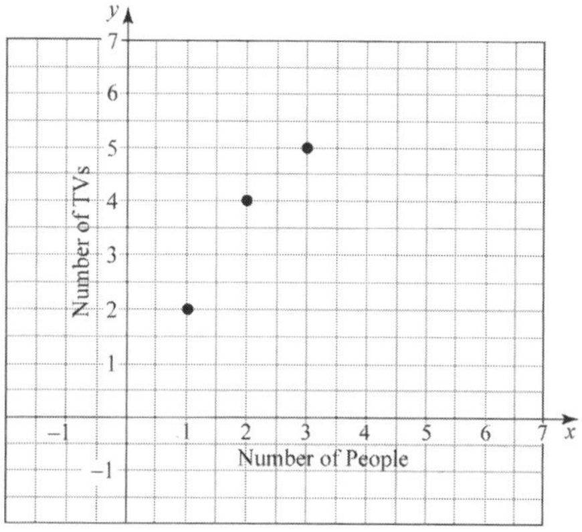

My Turn!

Plot the following paired data.

Number of Rooms in a House	Number of Paintings in a House
5	15
7	20
8	26

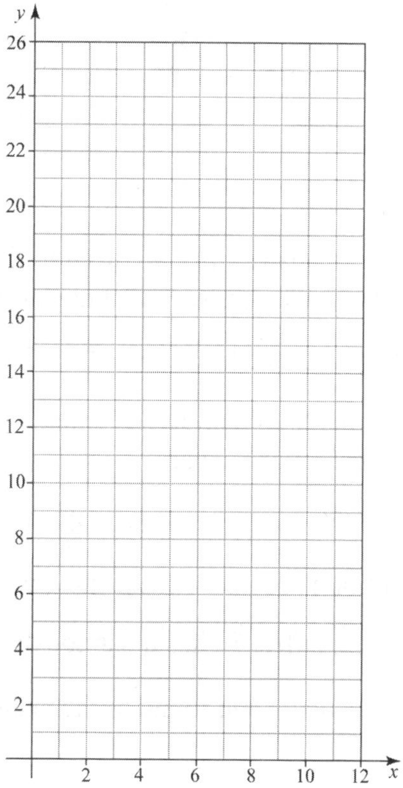

Objective 7: Review the skills for graphing.

In the corresponding chapter of the text, you will begin interpreting graphs. Some of the graphs will be in the Cartesian plane, as in the last example. Other graphs may include some that you may have seen before in the media, such as bar charts, pie charts, and time series graphs. We will do a sampling of two here. Notice how the graphs that follow are carefully labeled so that the information is clear to the reader. Whenever you create your own graphical representations of data, label everything clearly so that the graph conveys all the necessary information to the reader.

Example 12 The following is called a bar chart. It represents the types of patients that a veterinarian had in one day. What percentage of the vet's patients that day were cats? Round the answer to the nearest percent.

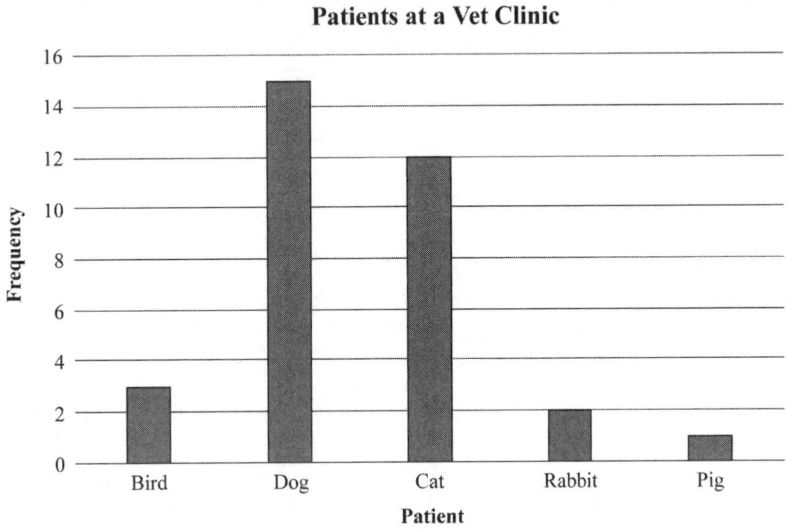

In order to answer this, we need to figure out how many cats were patients that day and how many animals were seen that day at the veterinarian's clinic.

The height of the bar labeled "Cat" will tell us that there were 12 cats seen that day.

To find the total number of animals seen that day, we have to sum up all of the heights (frequencies) of the bars.

Total number of animals seen that day $= 3+15+12+2+1 = 33$.

So, to find the relative frequency, we have to divide 12 by 33 to change it to a decimal: $\frac{12}{33} = 0.\overline{36}$. (The 36 has a bar above it since those digits are repeating.) We will now convert this to a percentage and round to the nearest percent.

$$0.\overline{36} = 36.\overline{36}\% \approx 36\%$$

Answer 36% of the patients were cats.

My Turn!

The bar chart represents the frequencies for the types of rooms that are in a home. What percentage of the house's rooms are bathrooms? Round the answer to the nearest percent.

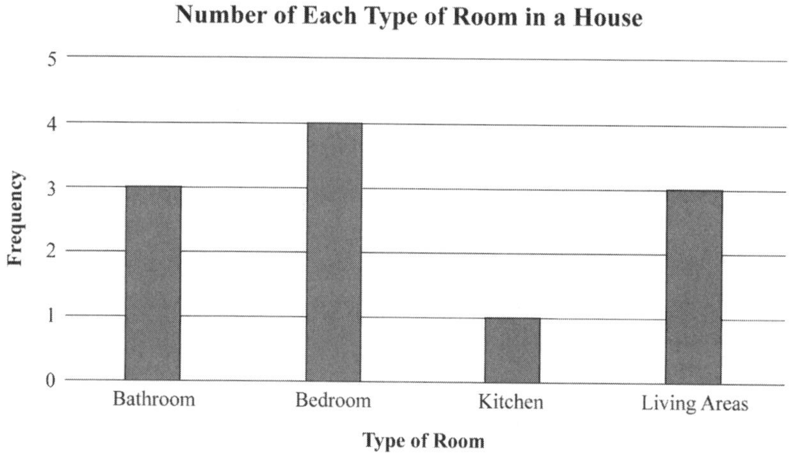

Example 13 The following is called a pie chart. It represents the college majors for a sample of students. What percentage of the college students were math majors? Round the answer to the nearest percent.

College Major

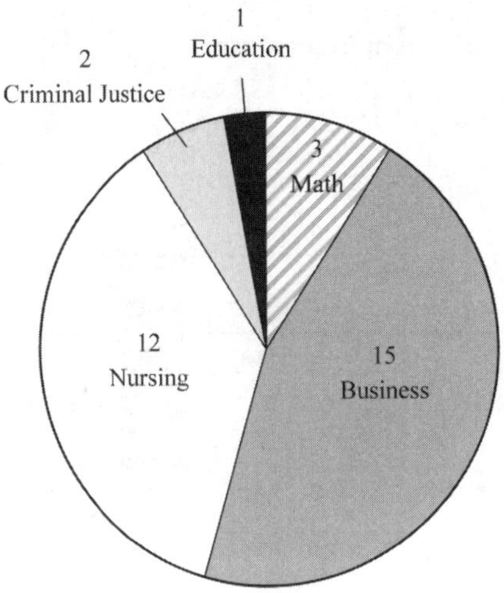

First, we need to figure out how many students have math as their college major. There were 3 math majors, as indicated by the number 3 in the appropriate slice of the pie chart. Next, we need to figure out the total number of students who were included in the sample. We can do that by summing up the frequencies for the 5 college majors.

Total number of students sampled $= 3+15+12+2+1=33$.

So in order to find the relative frequency, we have to divide 3 by 33 to change it to a decimal:
$\frac{3}{33} = 0.\overline{09}$.

We will now convert this to a percentage and round to the nearest percent.

$$0.\overline{09} = 9.\overline{09}\% \approx 9\%$$

Answer About 9% of college students sampled were math majors.

My Turn!

The pie chart represents the frequencies for various types of toys in a pediatrician's waiting room. What percentage of the toys were dolls? Round the answer to the nearest percent.

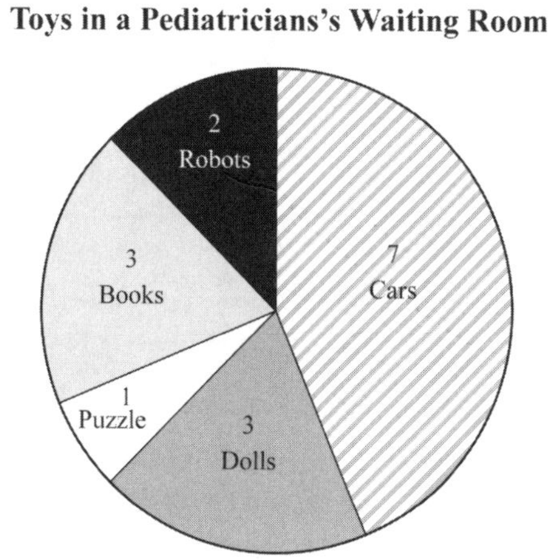

Answers to My Turn!

1. 7654.32 miles
2. 0.364 cm
3. $\dfrac{6}{17}$
4. 0.003
5. 230%
6. 56.25%
7. $\dfrac{7}{2500}$
8. 56.25%
9. 390
10. 24 respondents
11.

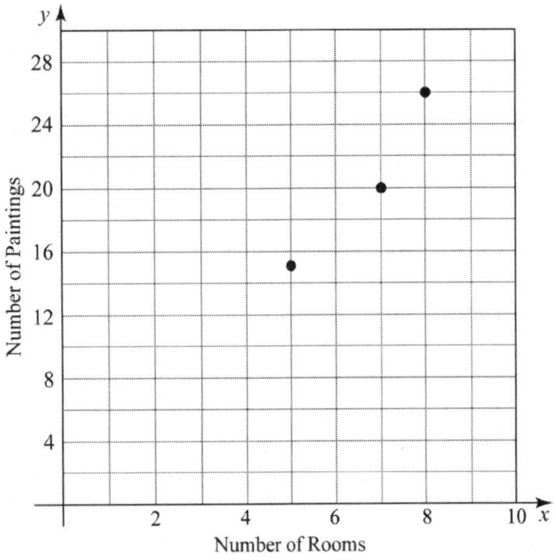

12. 27%
13. 19%

Practice Problems

1. Round 24,967.0784 kilograms to the nearest tenth of a kilogram.
2. Round 1.04708 pounds to 3 significant digits.
3. Write $\frac{84}{196}$ in lowest terms.
4. Convert 2.5% to a decimal.
5. Convert 0.007 to a percent.
6. Convert $\frac{2}{25}$ to a percent.
7. Convert 0.018 to a fraction.
8. If 30 out of 600 people tested positive for a cat allergy, calculate the percentage of participants who tested positive for a cat allergy.
9. What is 32% of 700?
10. A researcher surveyed 300 people. If 22% of those surveyed indicated that they are vegan, how many subjects responded that they are vegan?
11. Plot the following paired data.

Number of Times Exercising in a Week	Number of Meals Eaten in a Week
5	20
2	17
0	15

12. The bar chart represents the frequency of various colored suitcases on an airport baggage claim conveyor belt. What percentage of suitcases were red? Round the answer to the nearest percent.

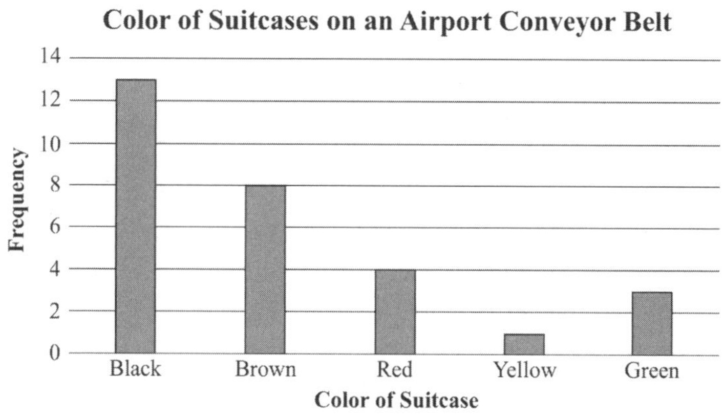

13. The pie chart represents the frequencies of various state license plates for cars on a Florida highway. What percentage of the license plates were from New York? Round the answer to the nearest percent.

License Plates for Cars on a Florida Highway

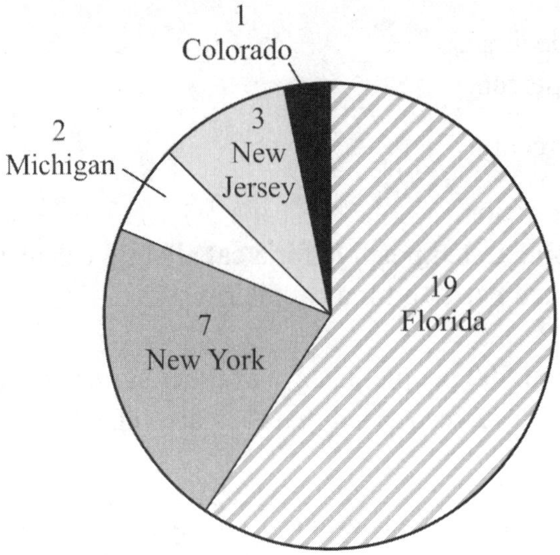

Integrated Review 3: Getting Ready to Evaluate Statistical Formulas

Elementary Statistics Chapter 3: Statistics for Describing, Exploring, and Comparing Data

Objectives:

1. Apply the order of operations.
2. Evaluate square roots.
3. Evaluate expressions and formulas.
4. Apply summation notation.

In this chapter of the Triola text, you will begin working with some basic statistical formulas that will be used to help describe data. In this integrated review chapter, we will review the concepts that will help you evaluate these statistical formulas.

Objective 1: Apply the order of operations.

We use the order of operations to simplify a numerical expression. You will have many numerical expressions that you will have to simplify in statistics, especially if you are evaluating formulas.

It is important that we do the operations in the correct order to arrive at a correct answer. It is critical that each step is completed from left to right.

Below you will find a listing of the order of operations. Always calculate the operations from left to right. This is especially important to remember for steps 3 and 4. This means that if there is a division to the left of multiplication in an expression, you would perform the division first.

1. **P**arentheses

2. **E**xponents

3. **M**ultiply or **D**ivide

4. **A**dd or **S**ubtract

Example 1 Perform the indicated operations: $5 \cdot 8^3$.

First we must calculate the exponent, since exponents should be done before multiplication according to the order of operations. (We reviewed calculating exponents in *Integrated Review* Chapter 1.)	$5 \cdot \boxed{8^3} = 5 \cdot \boxed{512}$
Next, we multiply.	$5 \boxed{\cdot} 8^3 = 5 \boxed{\cdot} 512 = 2560$

Answer $5 \cdot 8^3 = 2560$

My Turn!

Perform the indicated operations: $6^4 \div 4$.

Example 2 Perform the indicated operations: $15 \div 3 \cdot 2 + 9$.

First, we must multiply or divide from left to right. Since the division is to the left of the multiplication, we must do that first. It is a common misconception to think that you always multiply before dividing. This is not the case; you complete the operations of multiplication and division as they appear from left to right.	$15 \boxed{\div} 3 \cdot 2 + 9 = 5 \cdot 2 + 9$
Now, we multiply.	$5 \boxed{\cdot} 2 + 9 = 10 + 9$
Finally, we perform addition.	$10 \boxed{+} 9 = 19$

Answer $15 \div 3 \cdot 2 + 9 = 19$

My Turn!

Perform the indicated operations.

$5 + 2 \cdot 8$

Example 3: Perform the indicated operations: $\dfrac{(5-3)^2+(7-3)^2+(9-3)^2}{7}$

First, we must simplify within each set of parentheses.	$\dfrac{(5-3)^2+(7-3)^2+(9-3)^2}{7}=\dfrac{2^2+4^2+6^2}{7}$
Now, calculate the exponents.	$=\dfrac{4+16+36}{7}$
You can think of the entire numerator as being within an invisible set of parentheses. So, you would simplify that next.	$=\dfrac{56}{7}$
Finally, you can perform the division (or reduce the fraction).	$=8$

Answer $\dfrac{(5-3)^2+(7-3)^2+(9-3)^2}{7}=8$

My Turn!

Perform the indicated operations: $\dfrac{10}{(5+3)^2+(8+3)^2}$.

Round the answer to the nearest tenth.

Objective 2: Evaluate square roots.

Recall the meaning of square root.

The principal square root of a number a is denoted by $\sqrt{a}$. Furthermore, $x=\sqrt{a}$ if $x^2=a$. So, $3=\sqrt{9}$, since $3^2=9$.

Note that for the order of operations a square root can be treated as an exponent.

Example 4 Evaluate $\sqrt{64}$.

We are essentially trying to find a number such that the number times itself is 64.

$$\sqrt{64}=\sqrt{8\cdot 8}=8$$

Answer $\sqrt{64}=8$

My Turn!

Evaluate $\sqrt{25}$.

If the value under the radical symbol (i.e. the radicand) is not a perfect square, you are going to have to try a different strategy. You may recall having to simplify radicals for algebra. However, for the way that you will be using square roots for statistics, you won't have to simplify square roots as you would in an algebra course. You will simply use a calculator for radicands that are not perfect squares. However, you may want to come up with an estimate of which two integers the square root is between in order to catch careless calculator entry errors.

Example 5 Which two integers is $\sqrt{88}$ between?

You need figure out which two perfect squares 88 is between.

$$81<88<100$$

$$9^2<88<10^2$$

So,

$$9<\sqrt{88}<10$$

Answer: $\sqrt{88}$ is between 9 and 10.

My Turn!

Which two integers is $\sqrt{65}$ between?

Example 6 Approximate $\sqrt{88}$ to the nearest hundredth.

Use a calculator to evaluate this, since 88 is not a perfect square.

For the TI-83/84 calculators, the keystrokes are 2nd $\sqrt{\ }$ 88 enter. Note that the $\sqrt{\ }$ is the second function for the x^2 button. If you need to find other roots, press MATH.

Your calculator output is approximately 9.3808. This rounds to 9.38, which agrees with our findings from Example 5, as it is between 9 and 10.

Answer $\sqrt{88} \approx 9.38$. The symbol $\approx$ means "is approximately equal to." We use this symbol to indicate that our value is not an exact answer, since it is a rounded value.

My Turn!

Approximate $\sqrt{130}$ to the nearest hundredth.

Objective 3: Evaluate expressions and formulas.

Example 7 Perform the indicated operations:

$$\frac{18-11}{\frac{3}{\sqrt{9}}}$$

In order to simplify this expression, we will apply the order of operations to an expression that contains a square root. We treat a radical symbol as an exponent in the order of operations. We can do this because radicals can be rewritten as rational exponents (e.g. $\sqrt{9}=9^{\frac{1}{2}}$).

We can begin by viewing the numerator 18 – 11 as being contained within invisible parentheses and the denominator $\frac{3}{\sqrt{9}}$ as being contained within another set of parentheses. We can work on simplifying each of these. The simplification of the numerator is shown first.	$\frac{(18-11)}{\left(\frac{3}{\sqrt{9}}\right)}=\frac{7}{\left(\frac{3}{\sqrt{9}}\right)}$
To simplify the denominator $\frac{3}{\sqrt{9}}$, we must find $\sqrt{9}$.	$\frac{7}{\left(\frac{3}{\sqrt{9}}\right)}=\frac{7}{\left(\frac{3}{3}\right)}=\frac{7}{1}$
We find our final answer by dividing (or simplifying the fraction).	$\frac{7}{1}=7$

Answer $\frac{18-11}{\frac{3}{\sqrt{9}}}=7$

My Turn!

Perform the indicated operations:

$$\frac{6.8-6.1}{\frac{1.1}{\sqrt{16}}}$$

Round the answer to the nearest hundredth.

A formula is a special type of equation that expresses the relationship between several variables. On one side of the equation (usually the left side), you have the dependent variable. On the other side you have the independent variables. To evaluate a formula, we substitute known values for the dependent variables and follow the order of operations to simplify. We will include one example here, and you will see numerous examples in the chapters that follow.

Example 8 Evaluate the formula

$$E = z \cdot \frac{\sigma}{\sqrt{n}}$$

for $z = 1.96$, $\sigma = 2.1$, and $n = 100$. Round the answer to the nearest thousandth.

First, we must substitute the given values for the independent variables.

$$E = z \cdot \frac{\sigma}{\sqrt{n}} = 1.96 \cdot \frac{2.1}{\sqrt{100}}$$

In order to simplify the right-hand side, we will apply the order of operations to an expression that contains a square root.

We treat a radical symbol as an exponent in the order of operations. So, we will begin there. The square root of 100 is 10.	$E = 1.96 \cdot \frac{2.1}{\sqrt{100}} = 1.96 \cdot \frac{2.1}{10}$
We can simplify the fraction by dividing by 10.	$E = 1.96 \cdot \frac{2.1}{10} = 1.96 \cdot 0.21$
We find our value for the dependent variable, *E*, by multiplying.	$E = 1.96 \cdot 0.21 = 0.4116$
Now, we round our final answer to the nearest thousandth. Since the value in the ten thousandth place is a 6, we round to 0.412.	$E \approx 0.412$

Answer $E \approx 0.412$ when $E = z \cdot \frac{\sigma}{\sqrt{n}}$ and $z = 1.96$, $\sigma = 2.1$, and $n = 100$.

My Turn!

Evaluate the formula

$$n = \frac{z^2 \cdot \hat{p} \cdot \hat{q}}{E^2}$$

For $z = 2.575$, $\hat{p} = 0.3$, $\hat{q} = 0.7$, and $E = 0.05$. Round the answer to the nearest whole number.

Objective 4: Apply summation notation.

You will often see the summation symbol, Σ, in statistical formulas. It is the capital Greek letter "sigma." Whenever you see this symbol in a formula, it is telling you to add up values.

For instance, $\overline{x} = \dfrac{\Sigma x}{n}$ is the formula for finding the mean (average), $\overline{x}$, where x represents the data values and n is the number of data values.

Example 9 Find $\overline{x} = \dfrac{\Sigma x}{n}$ for the values in the accompanying table.

x
34
48
25
76
22
29

There are 6 values for x, so $n = 6$.

The formula is telling us to add up the values of x and divide by n.

$$\overline{x} = \frac{\Sigma x}{n} = \frac{34+48+25+76+22+29}{6} = \frac{234}{6} = 39$$

Note that the above formula could also be written as

$$\sum_{i=1}^{6} \frac{x_i}{n}$$

Answer $\overline{x} = 39$

My Turn!

Find $\overline{x} = \dfrac{\Sigma x}{n}$ for the values in the accompanying table, where n is the number of given values. Round to the nearest tenth, as needed.

x
5
4
8
7
2

Example 10 Find

$$\frac{\Sigma(x-39)^2}{n-1}$$

for the accompanying table and $n = 6$.

x
34
48
25
76
22
29

This notation is telling us to subtract 39 from each x value first, since we work out parentheses first in the order of operations. Next we would calculate exponents per the order of operations. Then, we sum up these squares of these differences. Once we have this sum, then we can perform division by the denominator.

$$\frac{\Sigma(x-39)^2}{n-1} = \frac{(34-39)^2+(48-39)^2+(25-39)^2+(76-39)^2+(22-39)^2+(29-39)^2}{6-1}$$

$$= \frac{(-5)^2+(9)^2+(-14)^2+(37)^2+(-17)^2+(-10)^2}{6-1}$$

$$= \frac{25+81+196+1369+289+100}{5}$$

$$= \frac{2060}{5}$$

$$= 412$$

Answer $\frac{\Sigma(x-39)^2}{n-1} = 412$ for the given table of values.

My Turn!

Find

$$\frac{\Sigma(x-2)^2}{n-1}$$

for the accompanying table and $n = 4$. Round to the nearest tenth, as needed.

x
4
3
2
1

Answer to My Turn!

1. 324
2. 21
3. 0.1
4. 5
5. 8 and 9
6. 11.4
7. 2.55
8. $n \approx 557$
9. $\overline{x} = 5.2$
10. 2

Practice Problems

1. Perform the indicated operations: $17-3\cdot 4$.
2. Perform the indicated operations: $5+3^4 \div 9$.
3. Perform the indicated operations: $\dfrac{4^2}{(9-2)^2+(8-2)^2}$. Round the answer to the nearest hundredth.
4. Evaluate $\sqrt{36}+\sqrt{100}$.
5. Which two integers is $\sqrt{101}$ between?
6. Approximate $\sqrt{255}$ to the nearest hundredth.
7. Perform the indicated operations. Round the answer to the nearest hundredth.
   $$\frac{0.9-0.8}{\sqrt{\dfrac{0.8\cdot 0.2}{50}}}$$
8. Evaluate the formula $\sigma=\sqrt{n\cdot p\cdot(1-p)}$ for $n=60$ and $p=0.25$. Round the answer to the nearest tenth.
9. Find $\dfrac{\Sigma x}{n}$ for the values in the given table, where n is the number of given values. Round to answer to the nearest tenth.
10. Find $\dfrac{\Sigma(x-2)^2}{n-1}$ for the values in the given table and $n=4$. Round the answer to the nearest tenth.

x
8
7
2
9

Integrated Review 4: Preparing for Probability

Elementary Statistics Chapter 4: Probability

Objectives:

1. Apply operations to fractions.
2. Apply operations to decimals.
3. Evaluate factorials.
4. Evaluate $_nP_r$ and $_nC_r$.
5. Determine the intersection, union, and complement of two sets.

In this chapter of the Triola text, you will learn the basics of probability. The probability, or likelihood, that some event will occur can be written either as a decimal or a fraction. For this reason, when you are evaluating probability formulas you will be dealing with operations on decimals and fractions. We will begin this integrated review chapter by reviewing operations on fractions and decimals. Then we will review some basic calculations that you will use for calculating other probability formulas. Finally, we will review the common operations of intersection, union, and complement of sets, as these concepts tie in with some other probability calculations. Once again, we will not explain the probability concepts in this integrated review. We will review only the basic calculation techniques here, and you will see the probability applications in the textbook itself.

Objective 1: Apply operations to fractions.

We will review the basic operations of addition, subtraction, multiplication, division, and exponentiation of fractions. Although you may typically calculate these using a calculator within a statistics course, we will briefly review how to do them without technology. However, you may also want to try each operation using a calculator to make sure you know how to work with fractions on your calculator.

Adding Fractions

Example 1 Perform the indicated operation:

$$\frac{3}{16}+\frac{9}{16}$$

In order to add fractions, you need a common denominator. The two given fractions already have the common denominator of 16. So, all you need to do is add the two numerators together and keep the common denominator. Always write your answers in reduced form.

$$\frac{3}{16}+\frac{9}{16}=\frac{(3+9)}{16}=\frac{12}{16}=\frac{3}{4}\cdot\frac{4}{4}=\frac{3}{4}\cdot 1=\frac{3}{4}$$

Answer $\frac{3}{16}+\frac{9}{16}=\frac{3}{4}$

My Turn!

Perform the indicated operation:

$$\frac{5}{8}+\frac{1}{8}$$

Example 2 Perform the indicated operation:

$$\frac{1}{9}+\frac{2}{15}$$

Since we need a common denominator in order to add or subtract fractions, our first step will be to rewrite each fraction over the lowest common denominator. The lowest common denominator is 45. Then, you follow the same procedure as in Example 1 and add the two numerators together and keep the common denominator.

$$\begin{aligned}\frac{1}{9}+\frac{2}{15}&=\frac{1}{9}\cdot\frac{5}{5}+\frac{2}{15}\cdot\frac{3}{3}\\&=\frac{5}{45}+\frac{6}{45}\\&=\frac{(5+6)}{45}\\&=\frac{11}{45}\end{aligned}$$

Answer $\frac{1}{9}+\frac{2}{15}=\frac{11}{45}$

My Turn!

Perform the indicated operation:

$$\frac{3}{10}+\frac{5}{12}$$

Subtracting Fractions

Example 3 Perform the indicated operation:

$$\frac{5}{6}-\frac{1}{3}$$

You also need a common denominator when subtracting fractions. So, our first step will be to rewrite each fraction over the lowest common denominator. The lowest common denominator is 6. Once you have rewritten each fraction over the lowest common denominator, subtract the two numerators and keep the common denominator. Be sure that your final answer is reduced!

$$\begin{aligned}\frac{5}{6}-\frac{1}{3}&=\frac{5}{6}-\frac{1}{3}\cdot\frac{2}{2}\\&=\frac{5}{6}-\frac{2}{6}\\&=\frac{(5-2)}{6}\\&=\frac{3}{6}\\&=\frac{1}{2}\end{aligned}$$

Answer $\frac{5}{6}-\frac{1}{3}=\frac{1}{2}$

My Turn!

Perform the indicated operation:

$$\frac{11}{15}-\frac{5}{12}$$

Since there are several topics in statistics that require subtracting fractions from 1, we will cover that type of subtraction problem here.

Example 4 Perform the indicated operation:

$$1-\frac{3}{8}$$

The number 1 can always be rewritten as $\frac{a}{a}, a \neq 0$. We will select a as the value of the common denominator. So, we set $a = 8$ for this problem. Then

$$1-\frac{3}{8}=\frac{8}{8}-\frac{3}{8}=\frac{5}{8}$$

Answer $1-\frac{3}{8}=\frac{5}{8}$

My Turn!

Perform the indicated operation:

$$1-\frac{16}{19}$$

Multiplying Fractions

Example 5 Perform the indicated operation:

$$\frac{5}{6}\cdot\frac{1}{3}$$

You do *not* need a common denominator in order to multiply fractions.

One way to multiply fractions is to multiply all numerators together and make that product the numerator, multiply all denominators together and make that product the denominator, and then reduce the fraction. Another way to do it is to "cancel out" any common factors that are in any numerator with any denominator (recall that you can do this because any number divided by itself is equal to 1) and then multiply across numerators and denominators. However, you do not have to decide on your preferred method for this particular example as there are no common factors in the numerators and denominators.

$$\frac{5}{6}\cdot\frac{1}{3}=\frac{5\cdot 1}{6\cdot 3}=\frac{5}{18}$$

Answer $\frac{5}{6}\cdot\frac{1}{3}=\frac{5}{18}$

My Turn!

Perform the indicated operation:

$$\frac{4}{15}\cdot\frac{5}{12}$$

Dividing Fractions

Example 6 Perform the indicated operation:

$$\frac{5}{6}\div\frac{1}{3}$$

You do *not* need a common denominator in order to divide fractions.

To divide, multiply by the reciprocal of the second fraction.

(Perhaps, you have heard the expression, “Keep. Change. Flip.” This pneumonic device helps some individuals remember to keep the first fraction the same, to change the division to multiplication, and to flip the divisor.)

Then, you follow the steps for multiplication. That is, multiply across the numerators, multiply across the denominators, and simplify.

$$\frac{5}{6}\div\frac{1}{3}=\frac{5}{6}\cdot\frac{3}{1}=\frac{5\cdot 3}{6\cdot 1}=\frac{15}{6}=\frac{5}{2}$$

You do not have to rewrite as a mixed number. Mixed numbers are generally not used in statistics.

Also, notice that you could have canceled out the common factor of 3 from the 3 in the second numerator and from the 6 in the first denominator before multiplying.

$$\frac{5}{6}\cdot\frac{3}{1}=\frac{5\cdot 1}{2\cdot 1}=\frac{5}{2}$$

Answer $\frac{5}{6}\div\frac{1}{3}=\frac{5}{2}$

My Turn!

Perform the indicated operation:

$$\frac{4}{9} \div \frac{7}{12}$$

Example 7 Perform the indicated operation:

$$\left(\frac{5}{6}\right)^3$$

Recall that an exponent is simply shorthand for repeated multiplication.

$$\left(\frac{5}{6}\right)^3 = \frac{5}{6} \cdot \frac{5}{6} \cdot \frac{5}{6} = \frac{125}{216}$$

Answer $\left(\frac{5}{6}\right)^3 = \frac{125}{216}$

My Turn!

Perform the indicated operation:

$$\left(\frac{2}{5}\right)^4$$

Objective 2: Apply operations to decimals.

We will review the basic operations of addition, subtraction, multiplication, division, and exponentiation of decimals. Although you may typically compute these using a calculator within a statistics course, we will briefly review how to do them without technology. Then, you may want to try these examples again using your calculator.

Adding Decimals

Example 8 Perform the indicated operation:

$$1.26+0.8$$

In order to add decimals, you need to line up the decimal points vertically. Then, add as you would normally and vertically line up the decimal point in the sum with those in the addends.

$$\begin{array}{r} 1.26 \\ \underline{+0.80} \\ 2.06 \end{array}$$

Answer $1.26+0.8=2.06$

My Turn!

Perform the indicated operation: $6.005+1.34$

Subtracting Decimals

Example 9 Perform the indicated operation:

$$1.26 - 0.8$$

You also need to line up the decimals points vertically when you are subtracting decimals.

$$\begin{array}{r} 1.26 \\ \underline{-0.80} \\ 0.46 \end{array}$$

Borrowing was involved in the subtraction. Since you will be using technology in this course, I encourage you to familiarize yourself with the procedure for performing basic operations on your preferred technology.

Answer $1.26 - 0.8 = 0.46$

My Turn!

Perform the indicated operation:

$$9 - 1.004$$

Multiplying Decimals

Example 10 Perform the indicated operation:

$$3.628 \cdot 0.2$$

You do *not* line up decimal points for multiplication.

Count the number of places to the right of each decimal point in each factor. After completing standard multiplication, you will have to insert the decimal point in the answer. The product will have the same number of digits to the right of the decimal point as the sum of the number of places to the right of the decimal point from the factors.

3.628 has three digits to the right of the decimal point.

0.2 has one digit to the right of the decimal point.

The product will have $3+1=4$ digits to the right of the decimal point.

$$\begin{array}{r} 3.628 \\ \times\ 0.2 \\ \hline 0.7256 \end{array}$$

Answer $3.628 \cdot 0.2 = 0.7256$

My Turn!

Perform the indicated operation:

$$5.2 \cdot 31.57$$

Dividing Decimals

Example 11 Perform the indicated operation:

$$3.628 \div 0.2$$

First, we will set up the values for long division.

$$0.2\overline{)3.628}$$

Then we must move the decimal point in the divisor so that it is at the right-hand side of the divisor. In this case, we would need to move it one place value to the right (equivalent to multiplying by 10). We also need to do this same movement to the decimal point in the dividend. (You can think about why we are allowed to move the decimal point the same number of places in the divisor and dividend.)

So, we now have

$$2\overline{)36.28}$$

Perform long division.

$$
\begin{array}{r@{}l}
 & 18.14 \\
2 & \overline{)36.28} \\
 & \underline{-2} \\
 & \ \ 16 \\
 & \underline{-16} \\
 & \ \ \ \ 02 \\
 & \ \ \ \underline{-2} \\
 & \ \ \ \ \ \ 08 \\
 & \ \ \ \ \ \underline{-8} \\
 & \ \ \ \ \ \ \ 0
\end{array}
$$

Make sure you line up the decimal point in your quotient directly above the one in the dividend.

Even though we have briefly reviewed the procedure here, you are strongly encouraged to calculate these operations with technology.

Answer $3.628 \div 0.2 = 18.14$

My Turn!

Perform the indicated operation:

$$34.6953 \div 5.1$$

Objective 3: Evaluate factorials.

Now that we have reviewed operations on fractions and decimals, let's go over some concepts that we use in counting and probability. The first notation that we will look at is called factorial notation. The symbol $n!$ is read as "*n* factorial" and is telling us to multiply the factors beginning with *n* all the way down to 1.

$$n! = n \cdot (n-1) \cdot (n-2) \cdot \ldots \cdot 3 \cdot 2 \cdot 1$$

Example 12 Evaluate $7!$

The $7!$ (7 factorial) is telling us to multiply the factors beginning with 7 all the way down to 1.

$$7! = 7 \cdot 6 \cdot 5 \cdot 4 \cdot 3 \cdot 2 \cdot 1 = 5040$$

Answer $7! = 5040$

My Turn!

Perform the indicated operation:

$$6!$$

Objective 4: Evaluate $_nP_r$ and $_nC_r$.

You will learn the contextual meaning of $_nP_r$ (permutation) within the Triola text. Here you will just practice the calculation.

$$_nP_r = \frac{n!}{(n-r)!}$$

Example 13 Evaluate $_{10}P_3$.

We begin by substituting 10 and 3 into the formula. Then, we will perform the subtraction within the parentheses in the denominator. If you think about it, this makes sense, since simplifying parentheses come first in the order of operations. We will then expand the factorials. Once this is written out, you will quickly see that many factors cancel out. Our answer is ultimately determined by multiplying together the three decreasing factors beginning with 10. Note that you are always able to compute $_nP_r$ by multiplying the *r* decreasing factors starting with *n*.

$$_{10}P_3 = \frac{10!}{(10-3)!} = \frac{10!}{7!} = \frac{10 \cdot 9 \cdot 8 \cdot 7 \cdot 6 \cdot 5 \cdot 4 \cdot 3 \cdot 2 \cdot 1}{7 \cdot 6 \cdot 5 \cdot 4 \cdot 3 \cdot 2 \cdot 1} = 10 \cdot 9 \cdot 8 = 720$$

Answer $_{10}P_3 = 720$

My Turn!

Evaluate ${}_8P_2$.

Now we will practice the calculating of combinations, which are denoted by ${}_nC_r$.

$$ {}_nC_r = \frac{n!}{r! \cdot (n-r)!} $$

Example 14 Evaluate ${}_{10}C_3$.

We begin by substituting 10 and 3 into the formula. Then, we will perform the subtraction within the parentheses in the denominator. Once this is written out, you will quickly see that many factors cancel out. Our answer is ultimately determined by multiplying together the three decreasing factors beginning with 10 and dividing by the three decreasing factors beginning with 3. Note that you are always able to compute ${}_nC_r$ by dividing the product of the r decreasing factors starting with n by $r!$.

$$ {}_{10}C_3 = \frac{10!}{3! \cdot (10-3)!} = \frac{10!}{3! \cdot 7!} = \frac{10 \cdot 9 \cdot 8 \cdot 7 \cdot 6 \cdot 5 \cdot 4 \cdot 3 \cdot 2 \cdot 1}{3 \cdot 2 \cdot 1 \cdot 7 \cdot 6 \cdot 5 \cdot 4 \cdot 3 \cdot 2 \cdot 1} = \frac{10 \cdot 9 \cdot 8}{3 \cdot 2 \cdot 1} = \frac{720}{6} = 120 $$

Answer ${}_{10}C_3 = 120$

My Turn!

Evaluate ${}_8C_2$.

Objective 5: Determine the intersection, union, and complement of two sets.

A set is simply a collection of items. We will review three operations on sets.

The **intersection** of two sets consists of all the elements that are simultaneously in both sets. The symbol that we use to indicate intersection is $\cap$.

Let A be a set and let B be a set; then $A \cap B$ consists of the elements that are in A **and** in B.

The **union** of two sets consists of all the elements that are in at least one of the sets. The symbol that we use to indicate union is $\cup$.

Let A be a set and let B be a set; then $A \cup B$ consists of the elements that are in A **or** in B.

The **complement** of a set A consists of the items that are in the universal set, U (the set that contains everything or all elements under consideration), that are **not** in A. Several symbols are commonly used to denote the complement; however, we will use $\overline{A}$ (read "A-bar").

Example 15 Let $U = \{1, 2, 3, 4, 5, 6, 7, 8, 9\}, A = \{2, 4, 6, 8\}$, and $B = \{1, 2, 3, 4, 5\}$.

a) Find $A \cap B$.
To find the intersection, we must look for the elements in A and in B.
$A = \{\boxed{2}, \boxed{4}, 6, 8\}$, and $B = \{1, \boxed{2}, 3, \boxed{4}, 5\}$ Those elements are 2 and 4.

$$A \cap B = \{2,4\}$$

b) Find $A \cup B$.
To find the union, we must look for the elements that are in A or in B (That is, join them all together). We only need to write each item once, though, in our solution set.

$$A \cup B = \{1,2,3,4,5,6,8\}$$

c) Find $\overline{A}$.

To find the complement of *A,* we must identify the elements in the universal set that are not in *A*. The elements in *A* are $A = \{2,4,6,8\}$. The elements that are in the universal set that are not in *A* are boxed: $U = \{\boxed{1},2,\boxed{3},4,\boxed{5},6,\boxed{7},8,\boxed{9}\}$. So, the complement of *A* consists of the elements 1, 3, 5, 7, and 9.

$$\overline{A} = \{1,3,5,7,9\}$$

Answer $A \cap B = \{2,4\}$, $A \cup B = \{1,2,3,4,5,6,8\}$, and $\overline{A} = \{1,3,5,7,9\}$

My Turn!

Let $U = \{1,2,3,4,5,6,7,8,9\}$, $A = \{1,2,3,4,5,6\}$, and $B = \{3,6\}$.

Find $A \cap B$, $A \cup B$, and $\overline{A}$.

Answers to My Turn!

1. $\frac{3}{4}$
2. $\frac{43}{60}$
3. $\frac{19}{60}$
4. $\frac{3}{19}$

5. $\frac{1}{9}$

6. $\frac{16}{21}$

7. $\frac{16}{625}$

8. 7.345
9. 7.996
10. 164.164
11. 6.803
12. 720
13. 56
14. 28
15. $A \cap B = \{3,6\}$, $A \cup B = \{1,2,3,4,5,6\}$, and $\overline{A} = \{7,8,9\}$

Practice Problems

1. Perform the indicated operation: $\frac{5}{18} + \frac{1}{18}$.

2. Perform the indicated operation: $\frac{3}{7} + \frac{7}{10}$.

3. Perform the indicated operation: $\frac{8}{15} - \frac{2}{5}$.

4. Perform the indicated operation: $1 - \frac{9}{10}$.

5. Perform the indicated operation: $\frac{3}{16} \cdot \frac{2}{9}$.

6. Perform the indicated operation: $\frac{4}{9} \div \frac{9}{16}$.

7. Perform the indicated operation: $\left(\frac{1}{4}\right)^3$.

8. Perform the indicated operation: $3.576 + 2.3$.

9. Perform the indicated operation: $5.6 - 2.354$.
10. Perform the indicated operation: $1.05 \cdot 2.5$.
11. Perform the indicated operation: $7.05 \div 0.5$.
12. Evaluate 9!.
13. Evaluate ${}_9P_4$.
14. Evaluate ${}_9C_4$.
15. Let $U =$ {red, blue, orange, yellow, green, purple, pink, white, black}, $A =$ {red, blue, yellow}, and $B =$ {red, pink, purple}.
    Find $A \cap B$, $A \cup B$, and $\overline{A}$.

Integrated Review 5: Evaluating Formulas for Probability Distributions

Elementary Statistics Chapter 5: Discrete Probability Distributions

Objectives:

1. Evaluate formulas for probability distributions.
2. Evaluate the binomial probability formula.
3. Evaluate expressions with *e*.
4. Evaluate the Poisson formula.

Chapter 5 of the Triola text will continue to discuss probability, and several related formulas will be introduced. In this integrated review chapter, we will go over how to evaluate some of those formulas. You will learn the meaning of the formulas in the corresponding chapter of the textbook.

Objective 1: Evaluate formulas for probability distributions.

We reviewed summation notation in *Integrated Review* Chapter 3. As you may recall, the summation symbol, Σ , tells you to add up values.

In the related chapter from the text, we will often check to see if the sum of a table of values equals 1.

Example 1 Determine whether $\Sigma P(x)=1$.

***P*(*x*)**
0.2
0.15
0.05
0.3
0.2
0.01

So, we will add up the six values that were given in the table and check to see if the sum is equal to 1.

$$\Sigma P(x)=0.2+0.15+0.05+0.3+0.2+0.01=0.91$$

$$0.91 \neq 1$$

Answer $\Sigma P(x)$ does not equal 1.

My Turn!

Determine whether $\Sigma P(x) = 1$.

***P*(*x*)**
0.03
0.22
0.15
0.2
0.25
0.15

Sometimes, the summation may require that you use values from two columns of a table.

Example 2 Find $\Sigma[x \cdot P(x)]$.

x	***P*(*x*)**
0	0.2
1	0.15
2	0.05
3	0.3
4	0.2
5	0.1

This notation is telling us to add up the products of the first column of *x* values with the second column, *P*(*x*). You can elect to find the sum in a horizontal fashion, or you can elect to insert a third column to the right with the products and then total that column.

$$\begin{aligned}\Sigma[x \cdot P(x)] &= (0 \cdot 0.2) + (1 \cdot 0.15) + (2 \cdot 0.05) + (3 \cdot 0.3) + (4 \cdot 0.2) + (5 \cdot 0.1) \\ &= 0 + 0.15 + 0.1 + 0.9 + 0.8 + 0.5 \\ &= 2.45\end{aligned}$$

Now, we will perform the same calculations in a table format.

x	$P(x)$	$x \cdot P(x)$
0	0.2	0
1	0.15	0.15
2	0.05	0.1
3	0.3	0.9
4	0.2	0.8
5	0.1	0.5
$\Sigma[x \cdot P(x)] =$		2.45

Answer $\Sigma[x \cdot P(x)] = 2.45$

My Turn!

Find $\Sigma[x \cdot P(x)]$.

x	$P(x)$
0	0.3
1	0.35
2	0.15
3	0.2

Example 3 Find $\Sigma\left[x^2 \cdot P(x)\right]$ using the same given table as in Example 2.

x	*P(x)*
0	0.2
1	0.15
2	0.05
3	0.3
4	0.2
5	0.1

This notation is telling us to add up the products of the squares of the first column of *x* values with the second column, $P(x)$. Again, this could be done horizontally or by using extra columns in a table. Note that we are following the order of operations and will be squaring before multiplying and then finally adding.

$$\begin{aligned}\Sigma\left[x^2 \cdot P(x)\right] &= (0^2 \cdot 0.2) + (1^2 \cdot 0.15) + (2^2 \cdot 0.05) + (3^2 \cdot 0.3) + (4^2 \cdot 0.2) + (5^2 \cdot 0.1) \\ &= (0 \cdot 0.2) + (1 \cdot 0.15) + (4 \cdot 0.05) + (9 \cdot 0.3) + (16 \cdot 0.2) + (25 \cdot 0.1) \\ &= 0 + 0.15 + 0.2 + 2.7 + 3.2 + 2.5 \\ &= 8.75\end{aligned}$$

Or, by expanding on the given table, we get the following:

x	x^2	*P(x)*	$x^2 \cdot P(x)$
0	0	0.2	0
1	1	0.15	0.15
2	4	0.05	0.2
3	9	0.3	2.7
4	16	0.2	3.2
5	25	0.1	0.25
	$\Sigma\left[x^2 \cdot P(x)\right] =$		8.75

Answer $\Sigma\left[x^2 \cdot P(x)\right] = 6.5$

My Turn!

Find $\Sigma\left[x^2 \cdot P(x)\right]$.

x	*P(x)*
0	0.3
1	0.35
2	0.15
3	0.2

Example 4 Evaluate $\mu + 2\sigma$ when $\mu = 25.2$ and $\sigma = 2.3$.

We need to substitute 25.2 for μ (the lowercase Greek letter mu) in the beginning of the expression and 1.3 for σ (the lowercase Greek letter sigma) at the end of the expression.

$\mu + 2\sigma = 25.2 + 2(2.3)$

Following the order of operations, we need to multiply before we add.

$25.2 + 2(2.3) = 25.2 + 4.6 = 29.8$

Answer $\mu + 2\sigma = 29.8$ when $\mu = 25.2$ and $\sigma = 2.3$.

My Turn!

Evaluate $\mu + 2\sigma$ when $\mu = 9.9$ and $\sigma = 0.4$.

Example 5 Evaluate $\mu = n \cdot p$ given $n = 150$ and $p = 0.35$.

We are evaluating a formula given the values for all but one variable, μ.

$$\mu = n \cdot p = 150 \cdot 0.35 = 52.5$$

Answer $\mu = 52.5$

My Turn!

Evaluate $\mu = n \cdot p$ when $n = 2000$ and $p = 0.04$.

Example 6 Evaluate $\sigma = \sqrt{np(1-p)}$ $n = 50$ and $n = 0.2$. Round the answer to the nearest tenth.

Begin by replacing *n* and *p* with the given values. You will be substituting 0.2 for *p* in two places. After you have done this, follow the order of operations.

$$\sigma = \sqrt{np(1-p)} = \sqrt{50 \cdot 0.2 \cdot (1-0.2)} = \sqrt{50 \cdot 0.2 \cdot (0.8)} = \sqrt{10 \cdot 0.8} = \sqrt{8} \approx 2.82 \approx 2.8$$

Answer $\sigma \approx 2.8$

My Turn!

Evaluate $\sigma = \sqrt{np(1-p)}$ for $n = 100$ and $p = 0.38$. Round the answer to the nearest tenth.

Objective 2: Evaluate the binomial probability formula.

Example 7 Evaluate ${}_nC_x \cdot p^x \cdot (1-p)^{n-x}$ for $n = 10$, $x = 2$, and $p = 0.3$. Round to three significant digits.

First, we must substitute the given values in for the given variables.

$${}_nC_x \cdot p^x \cdot (1-p)^{n-x} = {}_{10}C_2 \cdot 0.3^2 \cdot (1-0.3)^{10-2}$$

Next, we will clean up the third factor.

$$= {}_{10}C_2 \cdot 0.3^2 \cdot (0.7)^8$$

Then, we will evaluate the combination. Even though we are showing the detailed calculations here, you may want to practice evaluating combinations using technology.

$${}_{10}C_2 = \frac{10!}{2! \cdot (10-2)!} = \frac{10!}{2! \cdot 8!} = \frac{10 \cdot 9 \cdot 8 \cdot 7 \cdot 6 \cdot 5 \cdot 4 \cdot 3 \cdot 2 \cdot 1}{2 \cdot 1 \cdot 8 \cdot 7 \cdot 6 \cdot 5 \cdot 4 \cdot 3 \cdot 2 \cdot 1} = \frac{10 \cdot 9}{2 \cdot 1} = \frac{90}{2} = 45$$

We now have ${}_{10}C_2 \cdot 0.3^2 \cdot (0.7)^8 = 45 \cdot 0.3^2 \cdot (0.7)^8$. Per the order of operations, we will evaluate exponents next.

$$= 45 \cdot 0.09 \cdot 0.05764801$$

Next, we multiply.

$$= 0.2334744405$$

Finally, we round to 3 significant figures.

$$\approx 0.233$$

Answer ${}_nC_x \cdot p^x \cdot (1-p)^{n-x} \approx 0.233$ for $n = 10$, $x = 2$, and $p = 0.3$.

My Turn!

Evaluate ${}_nC_x \cdot p^x \cdot (1-p)^{n-x}$ for $n = 20$, $x = 5$, and $p = 0.4$. Round to 3 significant digits.

Objective 3: Evaluate expressions with *e*.

The mathematical constant *e* is the number $e \approx 2.71828$.

Example 8 Evaluate e^5. Round the answer to the nearest tenth.

You perform calculations with base *e* as you would with any other base. You will want to find powers of *e* using technology. (Most calculators have an e^x button.)

$$e^5 \approx 148.413 \approx 148.4$$

Answer $e^5 \approx 148.4$

My Turn!

Evaluate $3e^5$. Round the answer to the nearest tenth.

Objective 4: Evaluate the Poisson formula.

Example 9 Let $P(x)=\dfrac{\mu^x \cdot e^{-\mu}}{x!}$ and let $\mu = 2$. Find $P(3)$. Round your answer to three significant figures.

You may notice that this formula is written in function notation. We are being asked to evaluate the function when $x = 3$. So, we will substitute 2 in for μ and 3 in for *x*. Once we have done that, we will follow the order of operations. We will finish by rounding.

$$P(3)=\frac{2^3 \cdot e^{-2}}{3!} \approx \frac{8 \cdot 0.13533528}{3!} = \frac{1.08268224}{6} = 0.18044704 \approx 0.180$$

Answer $P(3) \approx 0.180$

My Turn!

Let $P(x)=\dfrac{\mu^x \cdot e^{-\mu}}{x!}$ and let $\mu = 4$. Find $P(3)$. Round your answer to 3 significant figures.

Answers to My Turn!

1. Yes
2. 1.25
3. 2.75
4. 10.7
5. 80
6. 4.9
7. 0.0746
8. 445.2
9. 0.195

Practice Problems

1. Determine whether $\Sigma P(x) = 1$.

***P*(*x*)**
0.18
0.02
0.35
0.2
0.05
0.15

2. Find $\Sigma[x \cdot P(x)]$.

x	***P*(*x*)**
0	0.2
1	0.45
2	0.15
3	0.2

3. Find $\Sigma[x^2 \cdot P(x)]$.

x	***P*(*x*)**
0	0.1
1	0.25
2	0.45
3	0.2

4. Evaluate $\mu - 2\sigma$ when $\mu = 265.9$ and $\sigma = 3.8$.
5. Evaluate $\mu = n \cdot p$ given $n = 1125$ and $p = 0.86$.
6. Evaluate $\sigma = \sqrt{np(1-p)}$ for $n = 200$ and $p = 0.48$. Round the answer to the nearest tenth.
7. Evaluate ${}_nC_x \cdot p^x \cdot (1-p)^{n-x}$ for $n = 15$, $x = 5$, and $p = 0.7$. Round to 3 significant digits.

8. Evaluate $6e^{-3}$. Round the answer to the nearest tenth.

9. Let $P(x)=\dfrac{\mu^x \cdot e^{-\mu}}{x!}$ and let $\mu=5$. Find $P(4)$. Round to 3 significant digits.

Integrated Review 6: The Building Blocks for Working with Normal Distributions

Elementary Statistics Chapter 6: Normal Probability Distributions

Objectives:

1. Find area.
2. Interpret inequality notation.
3. Evaluate formulas for normal probability distributions.

In this chapter of the Triola text, you will begin working with normal distributions (bell-shaped curves). You will be finding the area under portions of the normal curve. We will start this chapter of the integrated review by briefly looking at the area of a rectangle. Then, we will go over some of the basics for inequality notation. Finally, we will practice evaluating the formulas you will be using in the related chapter.

Objective 1: Find area.

In the chapter in the text on normal distributions, we will be finding area and associating it with probability. For the topic of uniform distributions, you will be finding area under rectangles. So, we will take a moment to review finding the area of a rectangle.

Example 1 Find the area of the rectangle shown.

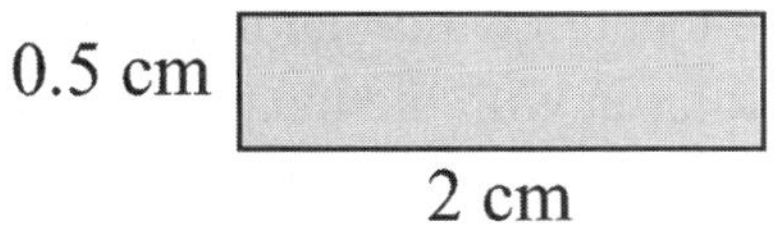

The area of a rectangle is given by $\text{base} \times \text{height}$ (or $\text{length} \times \text{width}$). We can write this as a formula.

$$A = b \cdot h$$

For the above rectangle, the base is 2 cm and the height is 0.5 cm.

We can plug these values into the formula.

$$A = b \cdot h = 2\,\text{cm} \cdot 0.5\,\text{cm} = 1\,\text{cm}^2$$

Answer The area is 1 cm^2.

My Turn!

Find the area of the shaded portion of the rectangle.

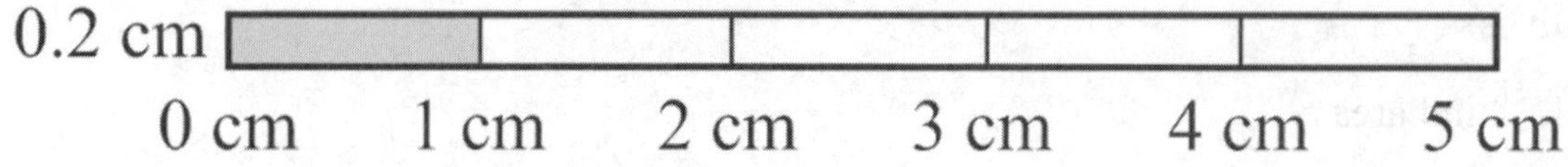

Objective 2: Interpret inequality notation.

In the related section of the textbook, we will be dealing with translating intervals in context to notation that we can use for calculating purposes. Also, you will want to be comfortable with indicating which integers would be located within an interval. The following table includes a column for inequality notation, a column with the inequality expressed in words, and a third column that indicates which integers from 0 to 10 satisfy the inequality from the first column.

Notation	**Translation**	**$X = \{0, 1, 2, 3, 4, 5, 6, 7, 8, 9, 10\}$**
$x < 5$	x is less than 5	0, 1, 2, 3, 4
$x > 7$	x is greater than 7	8, 9, 10
$2 < x < 6$	x is greater than 2 and less than 6	3, 4, 5
$x \leq 4$	x is less than or equal to 4	0, 1, 2, 3, 4
$x \geq 9$	x is greater than or equal to 9	9, 10

Example 2 Which integers from 0 to 10 satisfy the following inequality?

$$x < 7$$

This translates to "x is less than 7." The integers that are less than 7 include 0, 1, 2, 3, 4, 5, and 6.

Answer The integers from 0 to 10 that satisfy $x < 7$ are 0, 1, 2, 3, 4, 5, and 6.

My Turn!

Which integers from 5 to 15 satisfy the following inequality?

x is greater than or equal to 10.

There are some phrases that you may encounter as part of the text that will translate into inequalities. The following table includes a sampling of these.

Phrase	**Inequality**
At least 5 puppies	$x \geq 5$
No more than 5 puppies	$x \leq 5$
Fewer than 5 puppies	$x < 5$
More than 5 puppies	$x > 5$
Exactly 5 puppies	$x = 5$

There are other possibilities beyond the phrases used in the above table. The more problems you practice, the more comfortable you will be become with the various ways to describe inequalities.

Example 3 Translate the following phrases into inequalities.

a) A heart rate of at least 80 bpm

We can locate the phrase of *at least* in the table and see that it implies $\geq$. If a heart rate is at least 80 bpm, it means that it can be 80, 81, 82, …bpm.
Letting h represent heart rate, we can translate the phrase into $h \geq 80$.

b) A speed of no more than 55 mph

We can locate the phrase *no more than* in the second row of the table and see that it translates to $\leq$. If the speed is no more than 55 mph, it means that it can be anywhere from 0 mph up to and including 55 mph.
If we let s represent speed, the phrase can be rewritten as $s \leq 55$.

c) A temperature below 98.6 degrees Fahrenheit

The phrase *below* is not contained within the table. However, if we think about it, a temperature below 98.6 degrees means that it can be any value less than 98.6, not including 98.6.
Therefore, using t for temperature, our translation becomes $t < 98.6$.

Answer a) $h \geq 80$ b) $s \leq 55$ c) $t < 98.6$

My Turn!

Translate the following phrases into inequalities.

a) A weight of no more than 180 pounds.
b) A height of at least 60 inches
c) Exactly 9 doors in a house

Objective 3: Evaluating formulas for normal probability distributions.

You will be learning about several formulas related to the normal distribution in the corresponding section of the textbook. Here, we will practice evaluating these formulas. You will learn the meaning of the formulas and how to apply them in context within the text section itself.

Example 4 Evaluate $z=\dfrac{x-\mu}{\sigma}$ if $x = 21.3$, $\mu = 23.5$, and $\sigma = 2.1$. Round the answer to the nearest hundredth.

First, we must substitute for the variables in the correct location. Then, we must simplify the numerator. (This is because you can envision the entire numerator within parentheses, and the order of operations dictates that parentheses must be simplified first.) Next, we will divide, and then we will round our final answer.

$$z=\frac{21.3-23.5}{2.1}=\frac{-2.2}{2.1}\approx -1.047\approx -1.05$$

Answer $z \approx -1.05$

My Turn!

Evaluate $z=\dfrac{x-\mu}{\sigma}$ if $x = 10.3$, $\mu = 10.2$, and $\sigma = 1.3$. Round the answer to the nearest hundredth.

Example 5 Evaluate $x = \mu + z \cdot \sigma$ if $\mu = 6.7$, $z = 1.96$, and $\sigma = 0.9$. Round the answer to the nearest tenth.

First, we must substitute for the variables in the correct location. Then, we will follow the order of operations for the right-hand side. So, we will have to multiply before adding.

$$x = \mu + z \cdot \sigma = 6.7 + 1.96 \cdot 0.9 = 6.7 + 1.764 = 8.464 \approx 8.5$$

Answer $x \approx 8.5$

My Turn!

Evaluate $x = \mu + z \cdot \sigma$ if $\mu = 9.8$, $z = 2.575$, and $\sigma = 0.3$. Round the answer to the nearest tenth.

Example 6 Solve for x when $z = \dfrac{x - \mu}{\sigma}$ and $z = 2.15$, $\mu = 3.5$, and $\sigma = 0.3$. Round the answer to the nearest tenth.

We need to substitute 2.15 for z to the left of the equal sign, and 3.5 for μ and 0.3 for σ on the right-hand side.

$$2.15 = \frac{x - 3.5}{0.3}$$

Now, we need to solve for x. The first step is to multiply both sides of the equation by 0.3 to eliminate the fraction.

$$2.15(0.3) = \frac{(x - 3.5)}{0.3}(0.3)$$
$$0.645 = x - 3.5$$

Add 3.5 to both sides to isolate x.

$$0.645 + 3.5 = x - 3.5 + 3.5$$
$$4.145 = x$$

Now, we must round our final answer to the nearest tenth and $x \approx 4.1$.

Answer $x \approx 4.1$

My Turn!

Solve for x when $z = \dfrac{x - \mu}{\sigma}$ and $z = -0.34$, $\mu = 21.5$, and $\sigma = 0.8$. Round the answer to the nearest tenth.

Example 7 Evaluate

$$z = \frac{\overline{x} - \mu}{\dfrac{\sigma}{\sqrt{n}}}$$

if $\overline{x} = 9.8$, $\mu = 25.6$, $\sigma = 1.9$, and $n = 49$. Round your answer to the nearest hundredth.

We need to substitute 9.8 for $\overline{x}$ and 25.6 for μ in the numerator, and 1.9 for σ and 49 for n in the denominator.

$$z = \frac{9.8 - 25.6}{\frac{1.9}{\sqrt{49}}}$$

We can then simplify the right-hand side using the order of operations. You may want to visualize an invisible set of parentheses around the numerator and another invisible set of parentheses around the denominator.

$$\begin{aligned} z &= \frac{9.8 - 25.6}{\frac{1.9}{\sqrt{49}}} \\ &= \frac{(9.8 - 25.6)}{\left(\frac{1.9}{\sqrt{49}}\right)} \\ &= \frac{-15.8}{\left(\frac{1.9}{\sqrt{49}}\right)} \\ &= \frac{-15.8}{\left(\frac{1.9}{7}\right)} \\ &= \frac{-15.8}{0.27142857} \\ &= -58.2105266 \end{aligned}$$

Finally, we must round our final answer to the hundredth, which gives us $z \approx -58.21$.

Answer $z \approx -58.21$

My Turn!

Evaluate

$$z = \frac{\overline{x} - \mu}{\frac{\sigma}{\sqrt{n}}}$$

if $\overline{x} = 7.2, \mu = 5.8, \sigma = 1.2,$ and $n = 81$. Round the answer to the nearest hundredth.

Answers to My Turn!

1. 0.2 in^2
2. 10, 11, 12, 13, 14, 15
3. a) $w \leq 180$ b) $h \geq 60$ c) $d = 9$
4. $z \approx 0.08$
5. $x \approx 10.6$
6. $x \approx 21.2$
7. $z = 10.5$

Practice Problems

1. Find the area of the shaded region of the following rectangle.

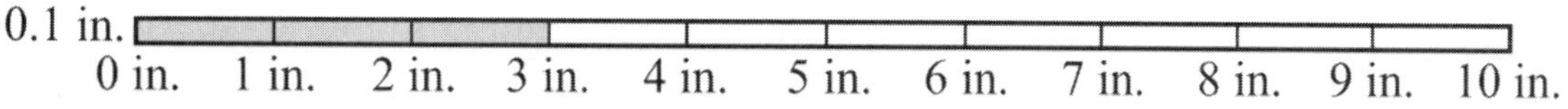

2. Which integers from 10 to 20 satisfy the following inequality? $x < 12$
3. Translate the following phrases into inequalities.
   A length of more than 18 inches
   A volume of less than 60 ft^3
4. Evaluate $z = \frac{x - \mu}{\sigma}$ if $x = 67.8$, $\mu = 70.2$, and $\sigma = 2.3$. Round the answer to the nearest hundredth.
5. Evaluate $x = \mu + z \cdot \sigma$ if $\mu = 78$, $z = 1.96$ and $\sigma = 10.3$. Round the answer to the nearest tenth
6. Solve for x when $z = \frac{x - \mu}{\sigma}$ and $z = 1.47$, $\mu = 99.5$ and $\sigma = 4.8$. Round the answer to the nearest tenth.
7. Evaluate
   $z = \frac{\overline{x} - \mu}{\frac{\sigma}{\sqrt{n}}}$ if $\overline{x} = 54.2, \mu = 52.6, \sigma = 3.2 \, and \, n = 64$. Round the answer to the nearest hundredth.

Integrated Review 7: Intervals of Numbers

Elementary Statistics Chapter 7: Estimating Parameters and Determining Sample Sizes

Objectives:

1. Find the middle value for an interval written either as an inequality or in interval notation.
2. Find the distance from the middle value of an interval to its endpoints.
3. Write and interpret three different forms of intervals (as used for confidence intervals).
4. Evaluate formulas used for confidence intervals.

In the chapter in the text on estimating parameters and determining sample sizes, we will be utilizing intervals a lot. So, in this integrated review chapter, we will spend some time discussing several topics related to intervals.

Objective 1: Find the middle value for an interval written either as an inequality or in interval notation.

One topic that you will have to understand in the text is how to find the middle value for a given interval. The given interval can be written either as an inequality or in interval notation. We will review how to handle both of these formats.

Example 1 Let $3.8 < \mu < 6.2$ represent an interval on the number line. Find the value that is in the middle of the interval.

From an algebra class, you may recall that this is the same thing as finding the midpoint for two points on a number line. In order to find the middle value, you have to add the two values of the endpoints and divide that sum by 2.

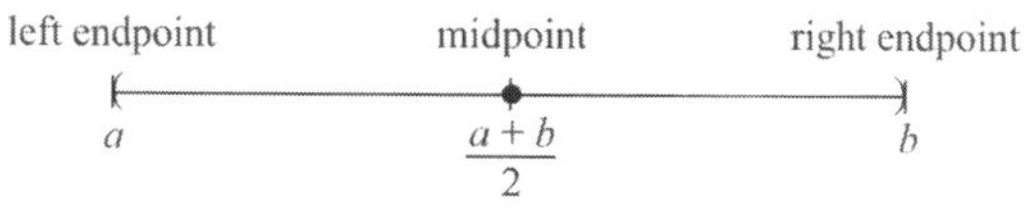

$$\text{The middle value of } a < \mu < b \text{ is } \frac{a+b}{2}.$$

We have $3.8 < \mu < 6.2$. So, substitute 3.8 for *a* and 6.2 for *b*. Then, the middle value of the given interval of numbers is found by following the order of operations. (Recall that the entire numerator can be thought of as being within parentheses and is therefore simplified first.)

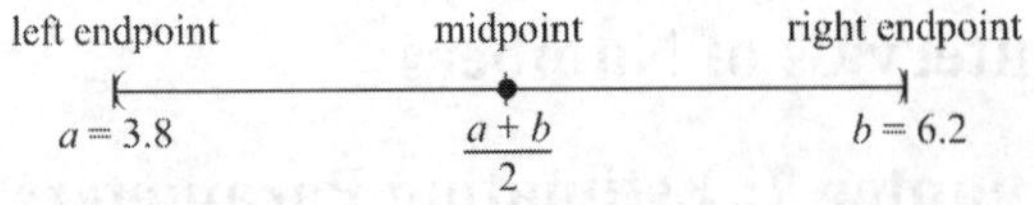

$$\frac{a+b}{2}=\frac{3.8+6.2}{2}=\frac{10}{2}=5$$

Answer The value of 5 is the middle value of the interval $3.8<\mu<6.2$.

My Turn!

Let $105.67<\mu<118.59$ represent an interval on the number line. Find the value that is in the middle of the interval.

Example 2 If $(0.124, 0.498)$ is an interval on the number line, find the value that is in the middle of the interval.

This problem is completed in the same way as Example 1. It simply differs in how the given interval is written. The interval $(0.124, 0.498)$ could be rewritten as $0.124<\mu<0.498$. In order to find the middle value, you have to add the two values of the endpoints and divide that sum by 2. That is,

$$\text{the middle value of } (a,b)=\frac{a+b}{2}$$

We have $(0.124, 0.498)=(a,b)$. So, substitute 0.124 for a and 0.498 for b and simplify.

$$\frac{a+b}{2}=\frac{0.124+0.498}{2}=\frac{0.622}{2}=0.311$$

Answer 0.311 is the middle value of the interval $(0.124, 0.498)$.

My Turn!

If $(6.48, 15.84)$ is an interval on the number line, find the value that is in the middle of the interval.

Objective 2: Find the distance from the middle value of an interval to its endpoints.

Example 3 Let $(0.368, 0.549)$ represent an interval on the number line. Find the distance from the middle of the interval to either endpoint.

Let E represent the distance from the midpoint to an endpoint.

There are two ways to solve this problem:

Method 1: Find the middle value of the interval first and then find the distance from the midpoint to the endpoint.

Step 1: Follow the process for Objective 1 to find the middle value of the interval.

The middle value of $(0.368, 0.549)$ is

$$\frac{0.368 + 0.549}{2} = \frac{0.917}{2} = 0.4585$$

Step 2: Find the distance from the middle value that you found in step 1 to the endpoint.

You can find the distance from the middle value to an endpoint by subtracting the left endpoint from the midpoint of the interval.

$$\text{distance from middle value to endpoint} = \text{middle value} - \text{left endpoint}$$

(You can also calculate this distance by subtracting the middle value from the right endpoint.)

For this example, the distance from the middle value to either endpoint is

$$E = 0.4585 - 0.368 = 0.0905$$

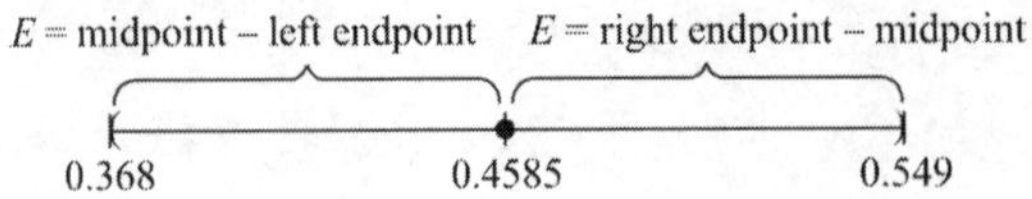

Method 2: Find the distance directly without actually finding the midpoint.

The distance from the middle value to the endpoint of an interval (a,b) can be given by

$$E = \frac{b-a}{2}$$

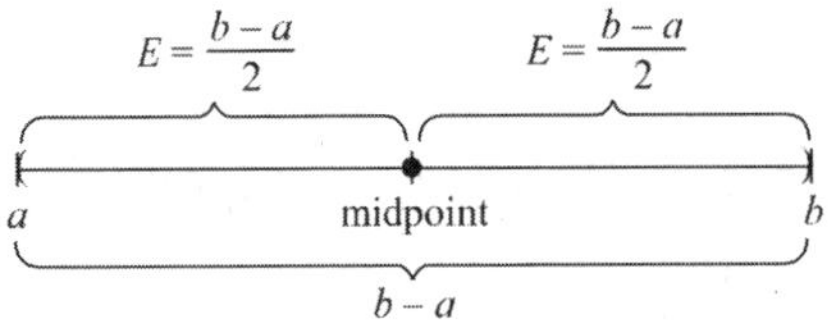

So, the distance from the middle value to the endpoint of the interval $(0.368,0.549)$ can be found as follows:

$$E = \frac{0.549 - 0.368}{2} = \frac{0.181}{2} = 0.0905$$

You can feel free to use the method that is more intuitive for you. Sometimes you may need to know the middle value anyway. In that case, method 1 is definitely the way to go. If you don't need to know the midpoint, then you may want to use method 2.

Answer The distance is 0.0905.

My Turn!

Let $(26.8, 32.4)$ represent an interval on the number line. Find the distance from the middle of the interval to either endpoint.

Objective 3: Write and interpret three different forms of intervals (as used for confidence intervals).

As you may have gathered from the examples above, there are several different forms of intervals. We will review the various forms you will see in the textbook. Often, technology will give you the output in a form that is different from how you may want to write it. So, it is important to understand how to translate from one format to another.

The key point that we have to remember is that $a < \mu < b$ and (a, b) represent the same interval for μ.

Example 4: Express $2.5 < \mu < 5.3$ in interval notation.

The endpoints of the interval are 2.5 and 5.3. So, we can rewrite $2.5 < \mu < 5.3$ as $(2.5, 5.3)$.

Answer $(2.5, 5.3)$

My Turn!

Express $-3.5 < \mu < 1.9$ in interval notation.

Example 5 Express $(0.88, 0.96)$ in inequality notation, assuming that the variable under study is μ.

The endpoints of the interval are 0.88 and 0.96. So, we can rewrite $(0.88, 0.96)$ as $0.88 < \mu < 0.96$.

Answer $0.88 < \mu < 0.96$

My Turn!

Express $(0.63, 0.92)$ in inequality notation, assuming that the variable under study is p.

Sometimes, you will want to write your interval in the following form:

$$\text{middle value}(m) \pm \text{distance from middle value to endpoint}(E) = m \pm E$$

The notation $\pm$ translates to *plus or minus*.

Example 6 Let $4.0 < \mu < 4.8$ represent an interval on the number line. Write the given interval in the format $m \pm E$.

First, we must find the value that is in the middle of the interval and let the variable m represent that.

$$m = (4.0, 4.8) = \frac{4.0 + 4.8}{2} = \frac{8.8}{2} = 4.4$$

Next, we must find the distance from the middle of the interval to either endpoint and let E represent that.

$$E = 4.4 - 4.0 = 0.4$$

Finally, we must write the given interval in the desired format $m \pm E$.

So, $4.0 < \mu < 4.8$ can be rewritten in the desired form of 4.4 ± 0.4.

Answer 4.4 ± 0.4

My Turn!

Let $0.18 < \mu < 0.36$ represent an interval on the number line. Write the given interval in the format $m \pm E$.

Objective 4: Evaluate formulas used for confidence intervals.

You will be learning about several formulas in the related section of the textbook on confidence intervals. Here, we will practice evaluating these formulas. You will learn the meaning of the formulas and how to apply them in context within the text section itself.

Example 7 Evaluate the formula

$$E = z \cdot \sqrt{\frac{\hat{p}(1-\hat{p})}{n}}$$

when $z = 1.96$, $\hat{p} = 0.516$, and $n = 1000$. Round the answer to the thousandths.

First, we must substitute for the variables in the correct location. Then, we will follow the order of operations for the right-hand side.

$$\begin{aligned}
E &= z \cdot \sqrt{\frac{\hat{p}(1-\hat{p})}{n}} \\
&= 1.96 \cdot \sqrt{\frac{0.516(1-0.516)}{1000}} \\
&= 1.96 \cdot \sqrt{\frac{0.516(0.484)}{1000}} \\
&= 1.96 \cdot \sqrt{\frac{0.249744}{1000}} \\
&= 1.96 \cdot \sqrt{0.000249744} \\
&\approx 1.96 \cdot 0.0158032908 \\
&= 0.03097445 \\
&\approx 0.0310
\end{aligned}$$

Answer $E \approx 0.0310$

My Turn!

Evaluate the formula $E = z \cdot \sqrt{\frac{\hat{p}(1-\hat{p})}{n}}$ when $z = 1.645$, $\hat{p} = 0.874$, and $n = 500$. Round the answer to the thousandths.

Example 8 Evaluate the formula

$$E = t \cdot \frac{s}{\sqrt{n}}$$

when $t = 2.626$, $s = 0.93$, and $n = 200$. Round your answer to the nearest hundredth.

By now, you probably have the idea that we need to substitute and simplify!

$$E = t \cdot \frac{s}{\sqrt{n}} = 2.626 \cdot \frac{0.93}{\sqrt{200}} \approx 2.626 \cdot \frac{0.93}{14.14213562} \approx 0.1726882039 \approx 0.17$$

Answer $E \approx 0.17$

My Turn!

Evaluate the formula

$$E = t \cdot \frac{s}{\sqrt{n}}$$

when $t = 2.756$, $s = 0.64$, and $n = 30$. Round your answer to the nearest hundredth.

Example 9 Evaluate the formula

$$n = \frac{z^2 \cdot p(1-p)}{E^2}$$

when $z = 1.645$, $p = 0.38$, and $E = 0.04$. Round the answer up to the next whole number.

Once again, substitute and simplify. If you are doing this on your calculator, be really careful with the order of operations.

$$\begin{aligned} n &= \frac{z^2 \cdot p(1-p)}{E^2} \\ &= \frac{1.645^2 \cdot 0.38(1-0.38)}{0.04^2} \\ &= \frac{1.645^2 \cdot 0.38(0.62)}{0.04^2} \\ &= \frac{2.706025 \cdot 0.38 \cdot 0.62}{0.0016} \\ &= \frac{0.63753949}{0.0016} \\ &\approx 398.4621813 \end{aligned}$$

Note the special rounding instructions for this problem. We are not following traditional rounding rules. If you don't get an exact whole number, you want to automatically go up to the next higher whole number. (You will find out why we do this when you learn about this formula in the textbook.)

So, for this problem we would round up to 399.

Answer $n = 399$

My Turn!

Evaluate the formula

$$n = \frac{z^2 \cdot p(1-p)}{E^2}$$

when $z = 1.96, p = 0.24$, and $E = 0.09$. Round the answer up to the next whole number.

Answers to My Turn!

1. 112.13
2. 11.16
3. 2.8
4. $(-3.5, 1.9)$
5. $0.63 < p < 0.92$
6. 0.27 ± 0.09
7. 0.024
8. 0.32
9. 87

Practice Problems

1. Let $0.45 < p < 0.85$ represent an interval on the number line. Find the value that is in the middle of the interval.
2. If $(2.57, 17.64)$ is an interval on the number line, find the value that is in the middle of the interval.
3. Let $(36.9, 42.4)$ represent an interval on the number line. Find the distance from the middle of the interval to either endpoint.
4. Express $3.5 < \mu < 11.9$ in interval notation.
5. Express $(0.63, 0.92)$ in inequality notation, assuming that the variable under study is μ.
6. Let $24.1 < \mu < 26.3$ represent an interval on the number line. Write the given interval in the format $m \pm E$.
7. Evaluate the formula

$$E = z \cdot \sqrt{\frac{\hat{p}(1-\hat{p})}{n}}$$

when $z = 1.96$, $\hat{p} = 0.526$, and $n = 300$. Round the answer to the thousandths.

8. Evaluate the formula

$$E = t \cdot \frac{s}{\sqrt{n}}$$

when $t = 2.678$, $s = 0.64$, and $n = 51$. Round your answer to the nearest hundredth.

9. Evaluate the formula

$$n = \frac{z^2 \cdot p(1-p)}{E^2}$$

when $z = 1.96, p = 0.37$, and $E = 0.08$. Round the answer up to the next whole number.

Integrated Review 8: Formulas for Hypothesis Testing

Elementary Statistics Chapter 8: Hypothesis Testing

Objective:

1. Evaluate formulas for hypothesis testing.

You will be learning about several formulas for hypothesis testing in the related section of the textbook. Here, we will practice evaluating these formulas. You will learn the meaning of the formulas and how to apply them in context within the text section itself.

Objective 1: Evaluate formulas for hypothesis testing.

Example 1 Evaluate the formula

$$z = \frac{\hat{p} - p}{\sqrt{\frac{p \cdot q}{n}}}$$

when $\hat{p} = \frac{x}{n}$, $x = 80$, $n = 100$, $p = 0.75$, and $q = 1 - p$. Round your answer to the nearest hundredth.

First we need to figure out the value for $\hat{p}$ so that we can substitute in that value. Since, $\hat{p} = \frac{x}{n}$, with $x = 80$ and $n = 100$, we get

$$\hat{p} = \frac{x}{n} = \frac{80}{100} = 0.8$$

We may also want to find the value of q. Since $q = 1 - p$ and $p = 0.75$ (Be really careful with p and $\hat{p}$, as they are indeed different), we get
$q = 1 - p = 1 - 0.75 = 0.25$

Now, we can substitute in values for all of the variables and solve for z.

$$z = \frac{\hat{p} - p}{\sqrt{\frac{p \cdot q}{n}}} = \frac{0.8 - 0.75}{\sqrt{\frac{0.75 \cdot 0.25}{100}}} = \frac{0.05}{\sqrt{\frac{0.1875}{100}}} = \frac{0.05}{\sqrt{0.001875}} \approx \frac{0.05}{0.0433} \approx 1.1547$$

Now, we will round our value for z to the nearest hundredth.

$$z \approx 1.15$$

Answer $z \approx 1.15$

My Turn!

Evaluate the formula

$$z = \frac{\hat{p} - p}{\sqrt{\frac{p \cdot q}{n}}}$$

when $\hat{p} = \frac{x}{n}$, $x = 90$, $n = 120$, $p = 0.35$ and $q = 1 - p$. Round your answer to the nearest hundredth.

Example 2 Evaluate the formula

$$t = \frac{\overline{x} - \mu}{\frac{s}{\sqrt{n}}}$$

when $\mu = 99.2$, $\overline{x} = 99.5$, $n = 100$, and $s = 1.3$. Round your answer to the nearest hundredth.

Begin by substituting the given values for the variables in the formula. Then, follow the order of operations to simplify the right hand side.

$$t = \frac{\overline{x} - \mu}{\frac{s}{\sqrt{n}}} = \frac{99.5 - 99.2}{\frac{1.3}{\sqrt{100}}} = \frac{0.3}{\frac{1.3}{\sqrt{100}}} = \frac{0.3}{\frac{1.3}{10}} = \frac{0.3}{0.13} \approx 2.3077 \approx 2.31$$

Note that if 100 were not a perfect square, you would want to use your calculator to find the value of the square root.

Answer $t \approx 2.31$

My Turn!

Evaluate the formula

$$t = \frac{\overline{x} - \mu}{\frac{s}{\sqrt{n}}}$$

when $\mu = 16$, $\overline{x} = 17.5$, $n = 80$, and $s = 2.4$. Round your answer to the nearest hundredth.

Example 3 Evaluate the formula

$$\chi^2 = \frac{(n-1) \cdot s^2}{\sigma^2}$$

when $\sigma = 3.9$, $n = 50$, and $s = 3.67$.

Round your answer to the nearest tenth.

Note that χ^2 (chi-squared) is considered as an entity. You do not approach this problem trying to solve for χ. You strive to get a value for χ^2. That is, $\chi^2 = \square$.

We can go straight to substituting the given values in for the variables and simplifying.

$$\begin{aligned}\chi^2 &= \frac{(n-1)\cdot s^2}{\sigma^2}\\ &= \frac{(50-1)\cdot 3.67^2}{3.9^2}\\ &= \frac{(49)\cdot 3.67^2}{3.9^2}\\ &= \frac{(49)\cdot 13.4689}{15.21}\\ &= \frac{659.9761}{15.21}\\ &\approx 43.3909\\ &\approx 43.4\end{aligned}$$

Answer $\chi^2 \approx 43.4$

My Turn!

Evaluate the formula

$$\chi^2 = \frac{(n-1)\cdot s^2}{\sigma^2}$$

when $\sigma = 4.5$, $n = 60$, and $s = 4.6$. Round your answer to the nearest tenth.

Answers to My Turn!

1. $z \approx 9.19$
2. $t \approx 5.59$
3. $\chi^2 \approx 61.7$

Practice Problems

1. Evaluate the formula

$$z = \frac{\hat{p} - p}{\sqrt{\frac{p \cdot q}{n}}}$$

when $\hat{p} = \frac{x}{n}$, $x = 90$, $n = 150$, $p = 0.82$, and $q = 1 - p$. Round your answer to the nearest hundredth.

2. Evaluate the formula

$$t = \frac{\overline{x} - \mu}{\frac{s}{\sqrt{n}}}$$

when $\mu = 44$, $\overline{x} = 45.2$, $n = 80$, and $s = 3.1$. Round your answer to the nearest hundredth.

3. Evaluate the formula

$$\chi^2 = \frac{(n-1) \cdot s^2}{\sigma^2}$$

when $\sigma = 11.3$, $n = 160$, and $s = 11.4$. Round your answer to the nearest tenth.

Integrated Review 9: Formulas for Two-Sample Hypothesis Testing

Elementary Statistics Chapter 9: Inferences from Two-Samples

Objective:

1. Evaluate formulas for two-sample hypothesis testing.

You will be learning more formulas for hypothesis testing in the related section of the textbook. Once again, we will practice evaluating these formulas here. You will learn the meaning of the formulas and how to apply them in context within the text section itself.

Objective 1: Evaluate formulas for two-sample hypothesis testing.

Example 1 Evaluate the formula

$$\bar{p} = \frac{x_1 + x_2}{n_1 + n_2}$$

for $x_1 = 15$, $x_2 = 18$, $n_1 = 35$, and $n_2 = 25$. Round your answer to the nearest ten thousandth.

First, we substitute the given values in for the variables in the formula. Notice that the variables with *x* and *n* have subscripts: The variables x_1 and x_2 represent two different unknowns. So, it is critical that we substitute the appropriate values into the variables with the correct subscripts.

$$\bar{p} = \frac{x_1 + x_2}{n_1 + n_2} = \frac{15+18}{35+25}$$

Once we have completed substituting, we apply the order of operations to the right-hand side. We can group the numerator and the denominator in two sets of parentheses to remind us to individually simplify each of them first by adding. Then, we divide the numerator by the denominator to change the fraction to a decimal. No rounding of the final answer is necessary in this problem.

$$\bar{p} = \frac{(15+18)}{(35+25)} = \frac{33}{60} = 0.55$$

Answer $\bar{p} = 0.55$

My Turn!

Evaluate the formula

$$\overline{p} = \frac{x_1 + x_2}{n_1 + n_2}$$

for $x_1 = 22$, $x_2 = 18$, $n_1 = 40$, and $n_2 = 30$. Round your answer to the nearest ten thousandth.

Example 2 Evaluate the formula

$$z = \frac{(\hat{p}_1 - \hat{p}_2) - (p_1 - p_2)}{\sqrt{\frac{\overline{p} \cdot \overline{q}}{n_1} + \frac{\overline{p} \cdot \overline{q}}{n_2}}}$$

if $p_1 - p_2 = 0$, $x_1 = 40$, $x_2 = 60$, $n_1 = 80$, $n_2 = 100$, $\hat{p}_1 = \frac{x_1}{n_1}$, $\hat{p}_2 = \frac{x_2}{n_2}$, $\overline{p} = \frac{x_1 + x_2}{n_1 + n_2}$, and $\overline{q} = 1 - \overline{p}$.

Round your answer to the nearest hundredth.

We have a lot of variables to substitute for in the formulas. We even have formulas to find out the values for some of these variables! Let's begin by using the formulas for the variables that we need to plug into the formula for *z*. We will find numerical values for $\hat{p}_1$, $\hat{p}_2$, $\overline{p}$, and $\overline{q} = 1 - \overline{p}$.

$$\hat{p}_1 = \frac{x_1}{n_1} = \frac{40}{80} = 0.5$$

$$\hat{p}_2 = \frac{x_2}{n_2} = \frac{60}{100} = 0.6$$

$$\overline{p} = \frac{x_1 + x_2}{n_1 + n_2} = \frac{40 + 60}{80 + 100} = \frac{100}{180} = 0.5556$$

$$\overline{q} = 1 - \overline{p} = 1 - 0.5556 = 0.4444$$

As you can see, we round $\overline{p}$ and $\overline{q}$ to the nearest ten thousandth even though we are rounding our final answer to the nearest hundredth. We keep a few extra places in intermediate steps to minimize rounding error. Ideally, we would try to complete all the calculations at one time using technology so that there is no intermediate rounding. However, when that is not possible, an extra two places should generally be sufficient.

Now, we are ready to substitute into the formula for z and simplify. Note that entire expression under the radical symbol can be thought of as being inside parentheses. This means we cannot take a square root until we have simplified the entire expression under the $\sqrt{}$.

$$
\begin{aligned}
z &= \frac{(\hat{p}_1 - \hat{p}_2) - (p_1 - p_2)}{\sqrt{\dfrac{\overline{p} \cdot \overline{q}}{n_1} + \dfrac{\overline{p} \cdot \overline{q}}{n_2}}} \\
&= \frac{(0.5 - 0.6) - 0}{\sqrt{\dfrac{0.5556 \cdot 0.4444}{80} + \dfrac{0.5556 \cdot 0.4444}{100}}} \\
&= \frac{-0.1}{\sqrt{\dfrac{0.5556 \cdot 0.4444}{80} + \dfrac{0.5556 \cdot 0.4444}{100}}} \\
&\approx \frac{-0.1}{\sqrt{\dfrac{0.2469}{80} + \dfrac{0.2469}{100}}} \\
&\approx \frac{-0.1}{\sqrt{0.0031 + 0.0025}} \\
&= \frac{-0.1}{\sqrt{0.0056}} \\
&\approx \frac{-0.1}{0.0748} \\
&\approx -1.3369 \\
&\approx -1.34
\end{aligned}
$$

Answer $z \approx -1.34$

My Turn!

Evaluate the formula

$$z = \frac{(\hat{p}_1 - \hat{p}_2) - (p_1 - p_2)}{\sqrt{\frac{\bar{p} \cdot \bar{q}}{n_1} + \frac{\bar{p} \cdot \bar{q}}{n_2}}}$$

if $p_1 - p_2 = 0$, $x_1 = 40$, $x_2 = 30$, $n_1 = 60$, $n_2 = 80$, $\hat{p}_1 = \frac{x_1}{n_1}$, $\hat{p}_2 = \frac{x_2}{n_2}$, $\bar{p} = \frac{x_1 + x_2}{n_1 + n_2}$, and $\bar{q} = 1 - \bar{p}$.

Round your answer to the nearest hundredth.

Example 3 Evaluate

$$E = z \cdot \sqrt{\frac{\hat{p}_1 \cdot \hat{q}_1}{n_1} + \frac{\hat{p}_2 \cdot \hat{q}_2}{n_2}}$$

if $z = 1.96$, $x_1 = 50$, $x_2 = 60$, $n_1 = 70$, $n_2 = 100$, $\hat{p}_1 = \frac{x_1}{n_1}$, $\hat{p}_2 = \frac{x_2}{n_2}$, $\hat{q}_1 = 1 - \hat{p}_1$, and $\hat{q}_2 = 1 - \hat{p}_2$.

Round your answer to 3 significant digits.

Let's begin by using the formulas for the variables that we need to plug into the formula for *E*. Let's find numerical values for $\hat{p}_1$, $\hat{p}_2$, $\hat{q}_1$, and $\hat{q}_2$. Throughout this solution, we will round to five significant digits (two more significant digits than what is desired in our final answer is normally adequate) in intermediate steps to minimize rounding error. Recall that the leading zeros don't count as significant digits.

$$\hat{p}_1 = \frac{x_1}{n_1} = \frac{50}{70} \approx 0.71429$$

$$\hat{p}_2 = \frac{x_2}{n_2} = \frac{60}{100} = 0.6$$

$$\hat{q}_1 = 1 - \hat{p}_1 \approx 1 - 0.71429 = 0.28571$$

$$\hat{q}_2 = 1 - \hat{p}_2 = 1 - 0.6 = 0.4$$

Since we now have numbers for each of the variables on the right-hand side, we can plug them into the formula for *E*. Then, we follow the order of operations on the right-hand side to simplify it.

$$\begin{aligned}
E &= z \cdot \sqrt{\frac{\hat{p}_1 \cdot \hat{q}_1}{n_1} + \frac{\hat{p}_2 \cdot \hat{q}_2}{n_2}} \\
&\approx 1.96\sqrt{\frac{0.71429 \cdot 0.28571}{70} + \frac{0.6 \cdot 0.4}{100}} \\
&\approx 1.96\sqrt{\frac{0.20408}{70} + \frac{0.24}{100}} \\
&\approx 1.96\sqrt{0.0029154 + 0.0024} \\
&= 1.96\sqrt{0.0053154} \\
&\approx 1.96 \cdot 0.072907 \\
&\approx 0.14290 \\
&\approx 0.143
\end{aligned}$$

Answer $E \approx 0.143$

My Turn!

Evaluate

$$E = z \cdot \sqrt{\frac{\hat{p}_1 \cdot \hat{q}_1}{n_1} + \frac{\hat{p}_2 \cdot \hat{q}_2}{n_2}}$$

if $z = 2.575$, $x_1 = 32$, $x_2 = 33, n_1 = 50, n_2 = 60,$ $\hat{p}_1 = \frac{x_1}{n_1}, \hat{p}_2 = \frac{x_2}{n_2},$ $\hat{q}_1 = 1 - \hat{p}_1$, and $\hat{q}_2 = 1 - \hat{p}_2$.

Round your answer to 3 significant digits.

Example 4 Evaluate the formula

$$t = \frac{(\bar{x}_1 - \bar{x}_2) - (\mu_1 - \mu_2)}{\sqrt{\frac{s_1^{\,2}}{n_1} + \frac{s_2^{\,2}}{n_2}}}$$

for $\bar{x}_1 = 210, \bar{x}_2 = 190,\ \mu_1 - \mu_2 = 0,\ s_1 = 11,\ s_2 = 7,\ n_1 = 45$, and $n_2 = 60$. Round your answer to the nearest hundredth.

We can begin this problem by immediately substituting the given values for the variables. Once again, we use the order of operations to guide us in simplifying the right-hand side. Be careful not to evaluate the square root until you have simplified the entire expression under the $\sqrt{\ }$.

$$\begin{aligned}
t &= \frac{(\bar{x}_1 - \bar{x}_2) - (\mu_1 - \mu_2)}{\sqrt{\frac{s_1^{\,2}}{n_1} + \frac{s_2^{\,2}}{n_2}}} \\
&= \frac{(210-190)-(0)}{\sqrt{\frac{11^2}{45} + \frac{7^2}{60}}} \\
&= \frac{20}{\sqrt{\frac{11^2}{45} + \frac{7^2}{60}}} \\
&= \frac{20}{\sqrt{\frac{121}{45} + \frac{49}{60}}} \\
&\approx \frac{20}{\sqrt{2.6889 + 0.8167}} \\
&= \frac{20}{\sqrt{3.5056}} \\
&\approx \frac{20}{1.8723} \\
&\approx 10.6820 \\
&\approx 10.68
\end{aligned}$$

Answer $t \approx 10.68$

My Turn!

Evaluate the formula

$$t = \frac{(\bar{x}_1 - \bar{x}_2) - (\mu_1 - \mu_2)}{\sqrt{\frac{s_1^2}{n_1} + \frac{s_2^2}{n_2}}}$$

for $\bar{x}_1 = 68, \bar{x}_2 = 64,\ \mu_1 - \mu_2 = 0,\ s_1 = 4,\ s_2 = 3,\ n_1 = 80$, and $n_2 = 90$.

Round your answer to the nearest hundredth.

Example 5 Evaluate

$$E = t \cdot \sqrt{\frac{s_1^2}{n_1} + \frac{s_2^2}{n_2}}$$

if $t = 2.449, s_1 = 1.5, s_2 = 1.8, n_1 = 50,$ and $n_2 = 40$. Round your answer to the nearest tenth.

By now, you know the routine; substitute and simplify!

$$\begin{aligned}
E &= t \cdot \sqrt{\frac{s_1^2}{n_1} + \frac{s_2^2}{n_2}} \\
&= 2.449 \cdot \sqrt{\frac{1.5^2}{50} + \frac{1.8^2}{40}} \\
&= 2.449 \cdot \sqrt{\frac{2.25}{50} + \frac{3.24}{40}} \\
&= 2.449 \cdot \sqrt{0.045 + 0.081} \\
&= 2.449 \cdot \sqrt{0.126} \\
&\approx 2.449 \cdot 0.3550 \\
&\approx 0.8694 \\
&\approx 0.9
\end{aligned}$$

Answer $E \approx 0.9$

My Turn!

Evaluate $E = t \cdot \sqrt{\frac{s_1^2}{n_1} + \frac{s_2^2}{n_2}}$ if $t = 1.99, s_1 = 0.6, s_2 = 0.8, n_1 = 30,$ and $n_2 = 50$. Round your answer to the nearest tenth.

Example 6 Evaluate

$$t = \frac{\bar{d} - \mu_d}{\frac{s_d}{\sqrt{n}}}$$

for $\bar{d} = 1.2, \mu_d = 0, s_d = 0.14,$ and $n = 49$. Round your answer to the nearest hundredth.

Let's substitute the four given values in for the appropriate variables. Be careful that you follow the order of operations carefully, especially for the denominator of the complex fraction. If you are using your calculator, you may have to insert parentheses.

$$t = \frac{\bar{d} - \mu_d}{\frac{s_d}{\sqrt{n}}} = \frac{1.2 - 0}{\frac{0.14}{\sqrt{49}}} = \frac{1.2}{\frac{0.14}{\sqrt{49}}} = \frac{1.2}{\frac{0.14}{7}} = \frac{1.2}{0.02} = 60$$

Answer $t = 60$

My Turn!

Evaluate

$$t = \frac{\bar{d} - \mu_d}{\frac{s_d}{\sqrt{n}}}$$

for $\bar{d} = 0.9, \mu_d = 0, s_d = 0.4,$ and $n = 64$. Round your answer to the nearest hundredth.

Answers to My Turn!

1. $\overline{p} \approx 0.5714$
2. $z \approx 3.42$
3. $E \approx 0.241$
4. $t \approx 7.30$
5. $E \approx 0.3$
6. $t = 18$

Practice Problems

1. Evaluate the formula

$$\overline{p} = \frac{x_1 + x_2}{n_1 + n_2}$$

for $x_1 = 46$, $x_2 = 37$, $n_1 = 80$, and $n_2 = 70$. Round your answer to the nearest ten thousandth.

2. Evaluate the formula

$$z = \frac{(\hat{p}_1 - \hat{p}_2) - (p_1 - p_2)}{\sqrt{\frac{\overline{p}\cdot\overline{q}}{n_1} + \frac{\overline{p}\cdot\overline{q}}{n_2}}}$$

if

$p_1 - p_2 = 0, x_1 = 96, x_2 = 83, n_1 = 100, n_2 = 100, \hat{p}_1 = \frac{x_1}{n_1}, \hat{p}_2 = \frac{x_2}{n_2}$, $\overline{p} = \frac{x_1 + x_2}{n_1 + n_2}$, and $\overline{q} = 1 - \overline{p}$.

Round your answer to the nearest hundredth.

3. Evaluate

$$E = z \cdot \sqrt{\frac{\hat{p}_1 \cdot \hat{q}_1}{n_1} + \frac{\hat{p}_2 \cdot \hat{q}_2}{n_2}}$$

if $z = 1.96$, $x_1 = 45$, $x_2 = 76, n_1 = 50, n_2 = 80$, $\hat{p}_1 = \frac{x_1}{n_1}, \hat{p}_2 = \frac{x_2}{n_2}$, $\hat{q}_1 = 1 - \hat{p}_1$, and

$\hat{q}_2 = 1 - \hat{p}_2$.

Round your answer to 3 significant digits.

4. Evaluate the formula

$$t = \frac{(\bar{x}_1 - \bar{x}_2) - (\mu_1 - \mu_2)}{\sqrt{\frac{s_1^2}{n_1} + \frac{s_2^2}{n_2}}}$$

for $\bar{x}_1 = 230, \bar{x}_2 = 242.5, \mu_1 - \mu_2 = 0, s_1 = 4.5, s_2 = 3.9, n_1 = 40,$ and $n_2 = 45$. Round your answer to the nearest hundredth.

5. Evaluate

$$E = t \cdot \sqrt{\frac{s_1^2}{n_1} + \frac{s_2^2}{n_2}}$$

if $t = 1.664, s_1 = 0.9, s_2 = 0.7, n_1 = 80,$ and $n_2 = 70$. Round your answer to the nearest tenth.

6. Evaluate

$$t = \frac{\bar{d} - \mu_d}{\frac{s_d}{\sqrt{n}}}$$

for $\bar{d} = 1.2, \mu_d = 0, s_d = 0.2,$ and $n = 100$. Round your answer to the nearest hundredth.

Integrated Review 10: Linear Equations

Elementary Statistics Chapter 10: Correlation and Regression

Objectives:

1. Find and interpret slope.
2. Find and interpret the y-intercept of a line.
3. Find values from a linear equation or graph.
4. Graph a linear equation.
5. Find and interpret a linear model ($y = mx + b$).

In this chapter of the Triola text, you will be modeling data with linear equations. In order to understand the statistical concepts, it is imperative that you are comfortable with lines in the coordinate plane. We will review slope, y-intercept, linear equations and their graphs, and linear modeling.

Objective 1: Find and interpret slope.

The slope of the line is its tilt or steepness.

Line with positive slope

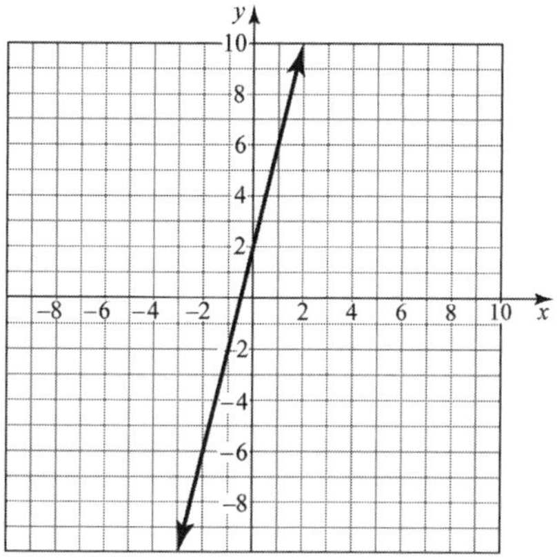

Line with negative slope

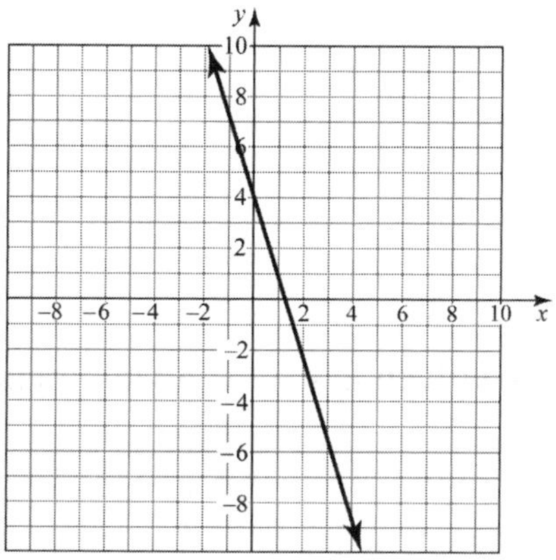

Line with slope of 0

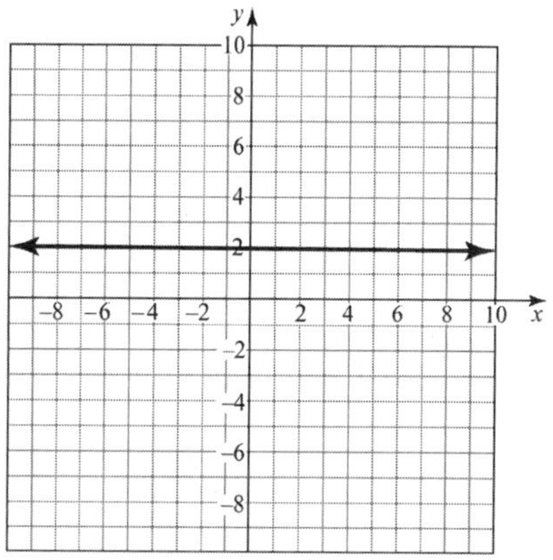

Here are some phrases related to slope that you may have heard:

$$\text{Slope} = \frac{\text{rise}}{\text{run}} = \frac{\text{change in y}}{\text{change in x}}$$

The slope of the line m between the points (x_1, y_1) and (x_2, y_2) can be given by

$$m = \frac{y_2 - y_1}{x_2 - x_1}$$

Example 1 Find the slope of the line that contains the points (3, 5) and (6, 9).

We can label the points as follows $(x_1, y_1) = (3,5)$ and $(x_2, y_2) = (6,9)$.

Now, we will substitute the given points into the formula for slope:

$$m = \frac{y_2 - y_1}{x_2 - x_1} = \frac{9-5}{6-3} = \frac{4}{3}$$

Answer The slope of the line that passes through the points (3, 5) and (6, 9) is $\frac{4}{3}$.

My Turn!

Find the slope of the line that contains the points (0.5, 2) and (0.7, 5). Round the answer to the nearest hundredth.

Example 2 Find the slope of the line given in the following graph.

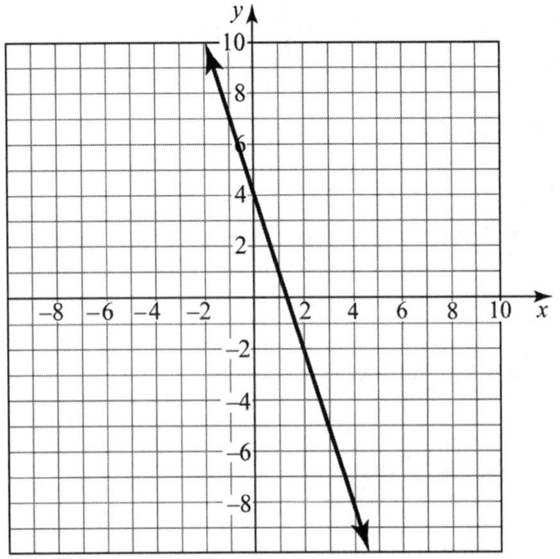

Begin by locating two points on the line. We will use the points (1, 1) and $(2,-2)$, although you could use any two points on the line.

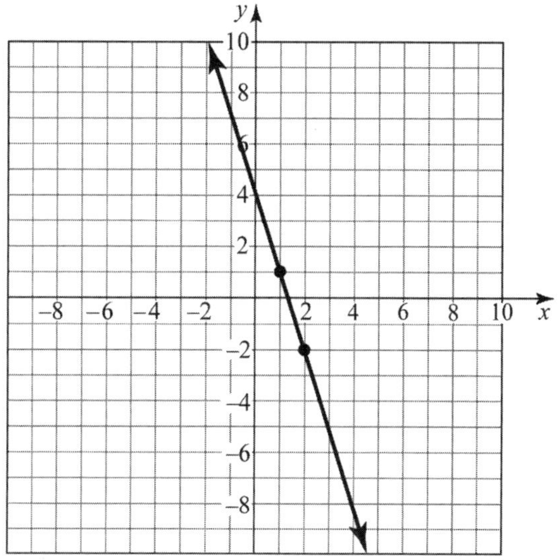

You could then either use the formula or map out the rise over run from the graph.

Formula Method:

$$m=\frac{y_2-y_1}{x_2-x_1}=\frac{-2-1}{2-1}=\frac{-3}{1}=-3$$

Rise-Over-Run Method: Let's look at this problem as if we are travelling from $(2,-2)$ to (1, 1). The rise from -2 to 1 is 3 units. The run from 2 to 1 is -1. The run is negative because we went to the left, which is the direction of the negatives on the number line. (You could have also traveled from (1, 1) to $(2,-2)$, which has a rise of -3 and a run of 1, resulting in the same slope.)

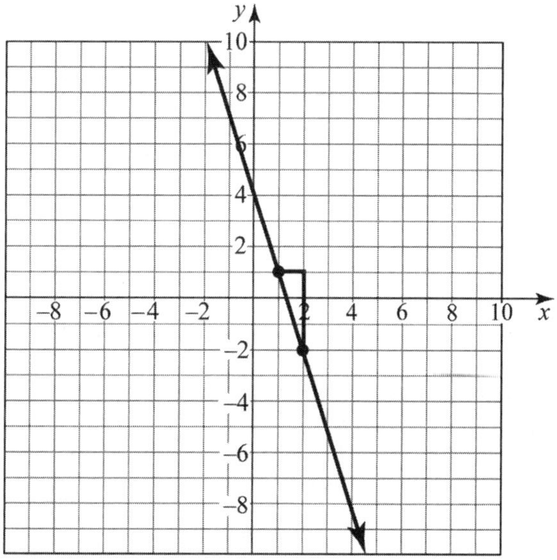

$$m=\frac{\text{rise}}{\text{run}}=\frac{3}{-1}=-3$$

Answer The slope is -3.

My Turn!

Find the slope of the line given in the following graph.

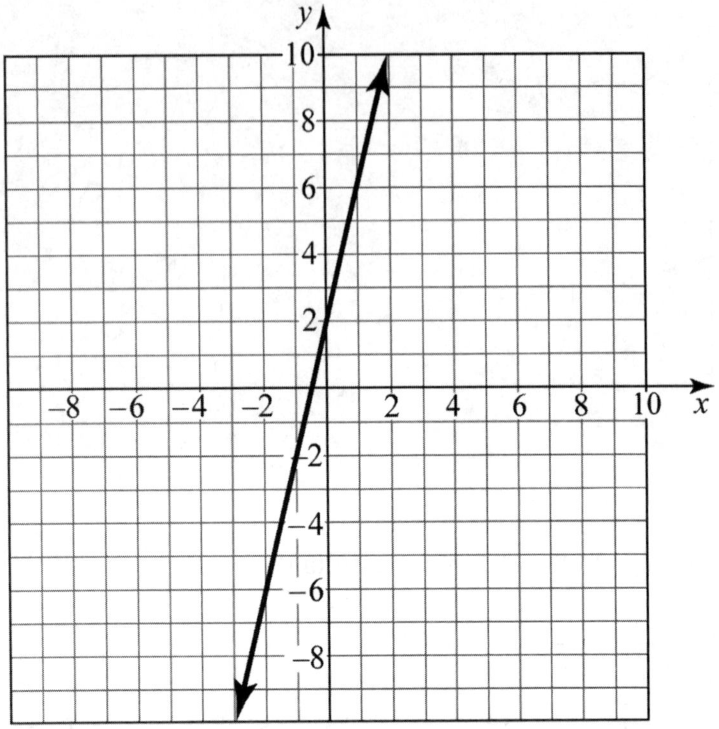

Objective 2: Find and interpret the y-intercept of a line.

The y-intercept(s) of a graph is (are) the point(s) where a graph crosses the y-axis. For a line, there is at most one y-intercept. Any y-intercept should be written as an ordered pair in the format $(0, b)$, where b is the y-coordinate of the y-intercept.

Example 3 Find the y-intercept of the following graph.

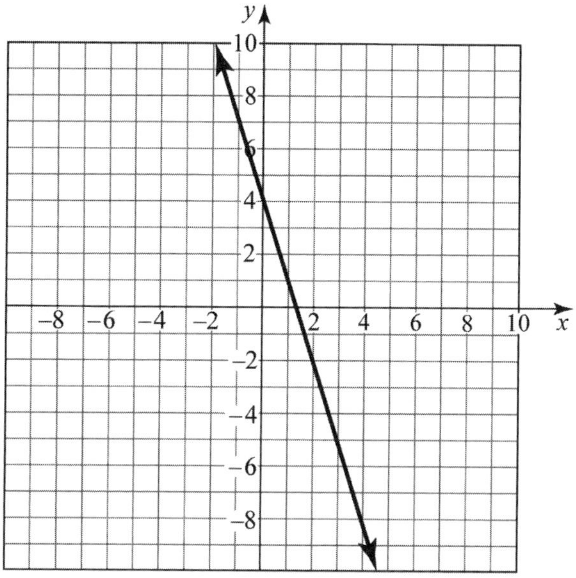

Looking at the graph, we can see that the line clearly crosses through the y-axis at the point (0,4).

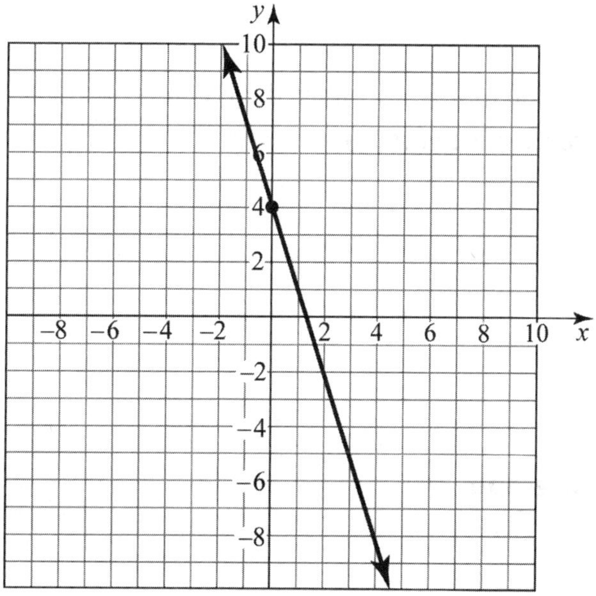

Answer The y-intercept of the line is (0, 4).

My Turn!

Find the y-intercept of the following graph.

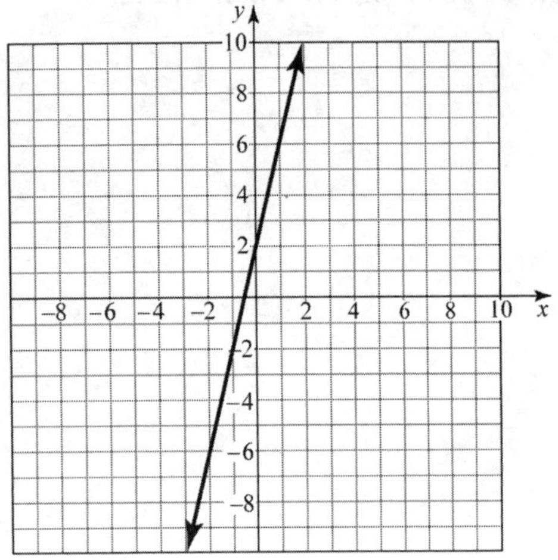

For any linear equation written in slope-intercept form, $y = mx + b$. The y-intercept can be written as $(0, b)$. That is, the constant value b is the y-coordinate of the y-intercept.

Example 4 Find the y-intercept of the line $y = 7x - 3$.

The equation is written in the form $y = mx + b$. The value of b is -3. Since the y-intercept should be written as an ordered pair $(0, b)$, the y-intercept for this line is $(0, -3)$.

Answer The y-intercept is $(0, -3)$.

My Turn!

Find the y-intercept of the line.

$$\hat{y} = \frac{2}{3}x + \frac{1}{2}$$

Objective 3: Find values from a linear equation or graph.

Example 5 Find three pairs of data values that would be on the line $\hat{y} = 5x + 7$.

There are actually an infinite number of points on any line. So, answers can vary.

If you plug in any x value, you will get a corresponding $\hat{y}$ (y-hat) value.

Let's plug in 0, 1, and 2 for x.

x	$\hat{y} = 5x + 7$	**Paired data values**
0	$\hat{y}=5(0)+7=0+7=7$	(0, 7)
1	$\hat{y}=5(1)+7=5+7=12$	(1, 12)
2	$\hat{y}=5(2)+7=10+7=17$	(2, 17)

Answer The points (0, 7), (1, 12), and (2, 17) are on the line $\hat{y} = 5x + 7$.

My Turn!

Find three pairs of data values that would be on the line $\hat{y} = -2x + 3$.

Example 6 In the following graph, the horizontal axis represents the height of human beings and the vertical axis represents the weight of human beings. Find the weight (in pounds) for somebody who is 60 inches tall.

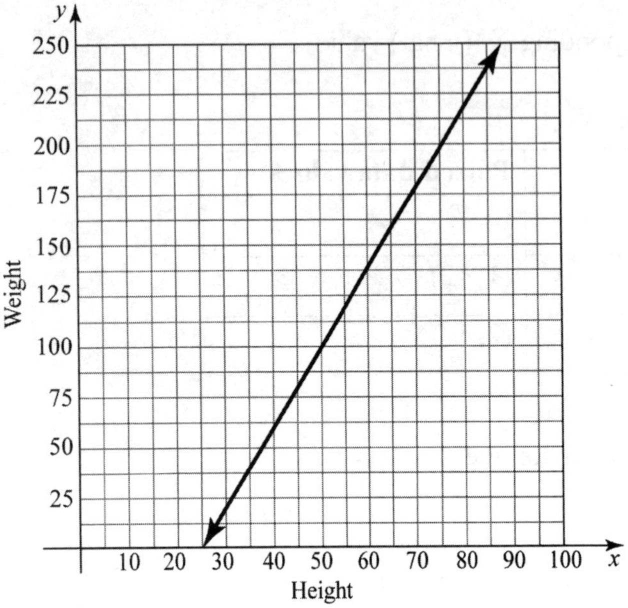

The horizontal axis represents height and the vertical axis represents weight. So, we begin by locating the given value of height = 60 on the horizontal axis and then move up to locate the point on the given line. Then, we look over to the vertical axis and find the corresponding y value.

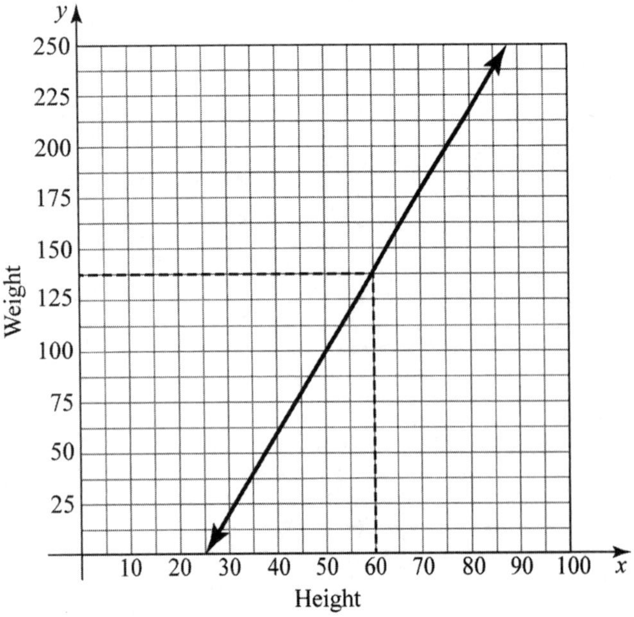

The corresponding weight for a height of 60 inches looks to be about 137.5 pounds, since it halfway between 125 and 150 pounds.

Answer 137.5 pounds

My Turn!

In the following graph, the horizontal axis represents the weight (in pounds) of dogs and the vertical axis represents the weight (in pounds) that the dog eats in a week. How many pounds of dog food does a dog who weighs 48 pounds eat in a week?

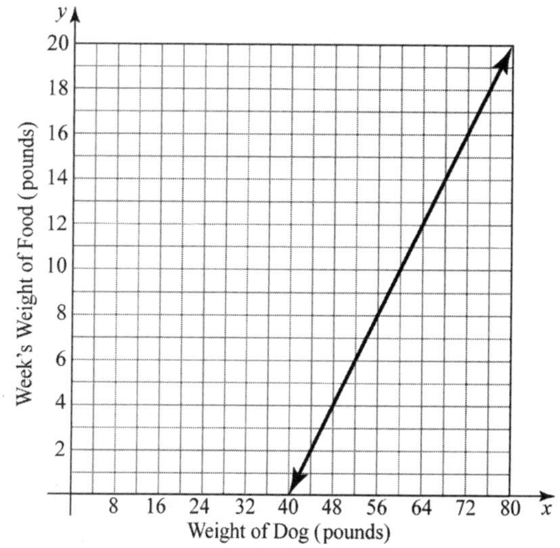

Objective 4: Graph a linear equation.

Example 7 Graph $y = -\frac{2}{3}x + 5$.

There are quite a few strategies for graphing lines from an equation. For this example, we will use the slope and the y-intercept to graph the line.

First, extract the y-intercept from the linear equation. Since $b = 5$, the y-intercept for this line is (0, 5). Plot the y-intercept.

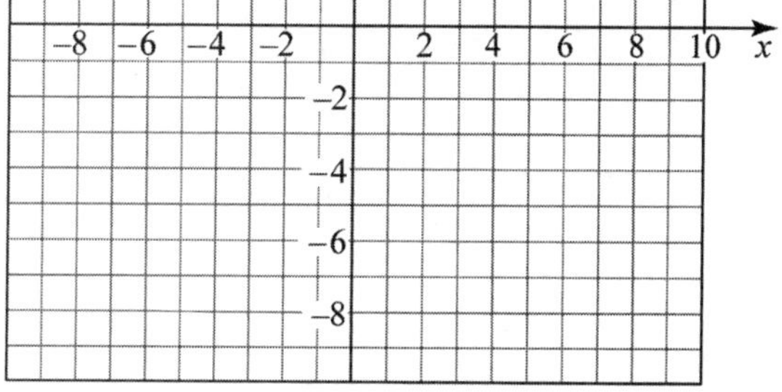

Once you have the y-intercept, you can use the slope to locate another point. The slope in this problem is $-\frac{2}{3}$. So, another point will be down 2 units and to the right 3 units. (Remember rise/run.)

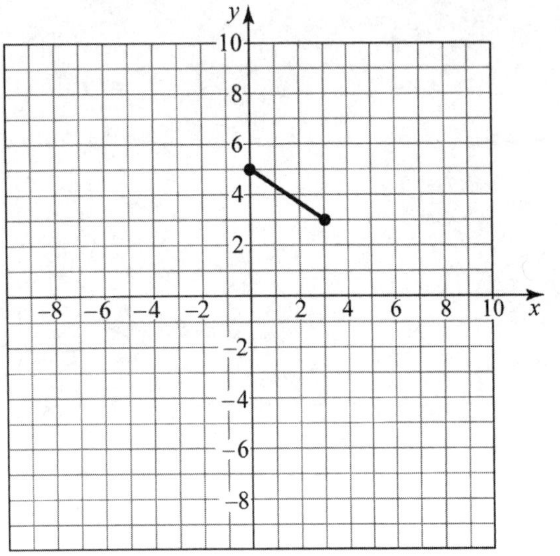

You can now extend this line using a straightedge.

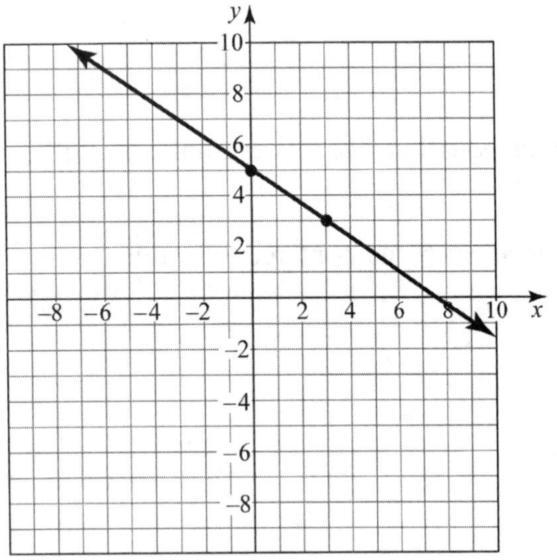

My Turn!

Graph $y = \frac{1}{4}x + 3$.

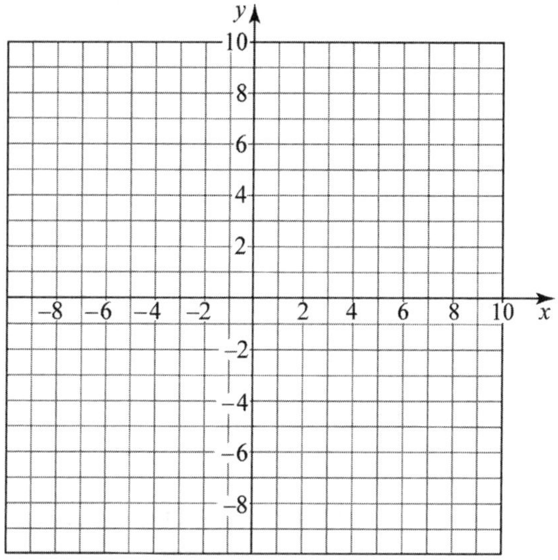

Example 8 Graph $y = 4x - 10$.

You could graph this equation using the method of Example 6 by using the *y*-intercept of $(0, -10)$ and the slope of 4. However, we will review another method in this example. You can find two points on the line and then connect them with a straightedge. Finding a third point is recommended as a check (or you could use the *y*-intercept as a check).

x	$y = 4x - 10$	**Paired data values**
1	$y = 4(1) - 10 = 4 - 10 = -6$	$(1, -6)$
2	$y = 4(2) - 10 = 8 - 10 = -2$	$(2, -2)$

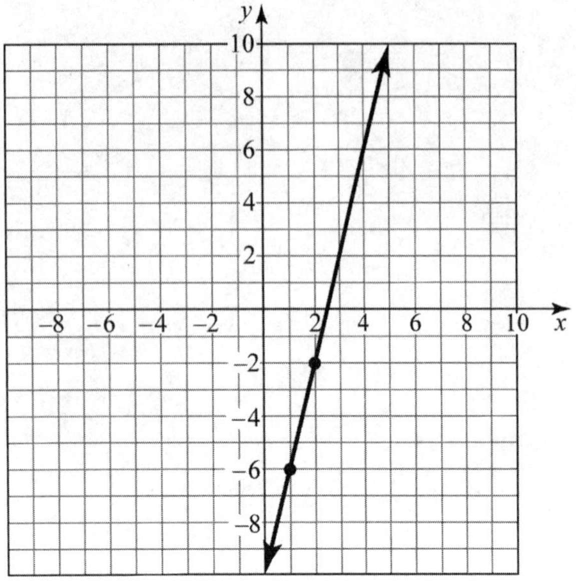

We can see that the line also passes through the y-intercept $(0,-10)$, which serves as our check.

My Turn!

Graph $y=\frac{1}{2}x-5$. (Hint: You may want to substitute values for x that are multiples of the denominator.)

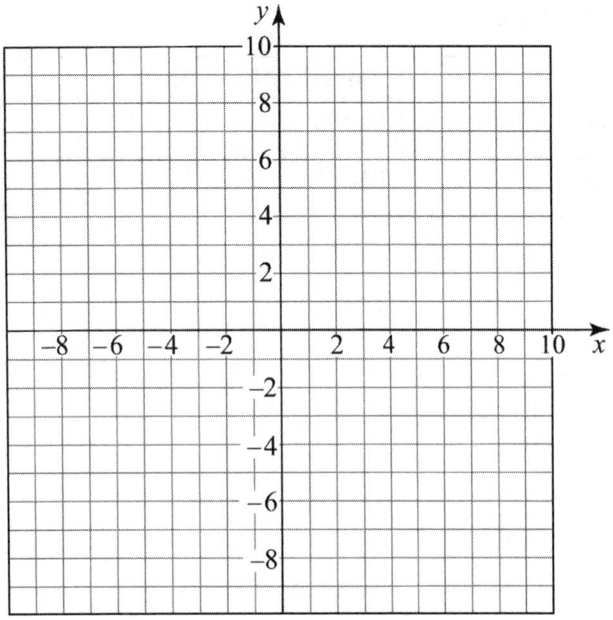

Objective 5: Find and interpret a linear model ($\hat{y} = mx + b$).

You will be studying linear models in great detail in the textbook. For those problems, you will be looking at real data that can sometimes be messy and may best be handled with a calculator. In this workbook, we will use a simplistic model to help prepare you.

Example 9 Find a linear model for the following data for the number of pages in a book and its weight in ounces.

Number of Pages	Weight (oz.)
150	6
200	8
250	10
300	12

The first thing you may want to do is plot the points to determine whether they fall on a line.

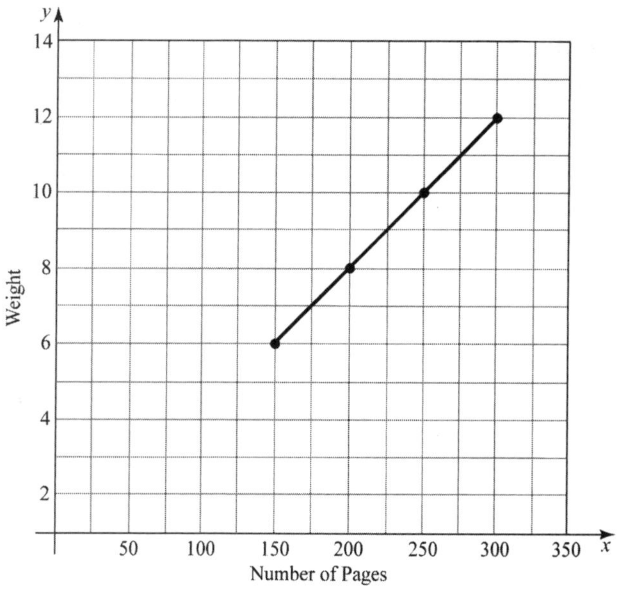

It does appear that the data fall on a line. So, we will use two points to find the slope.

$$m = \frac{8-6}{200-150} = \frac{2}{50} = \frac{1}{25}$$

We now need to find the y- intercept. Since it was not one of the given points, we can substitute any given point for x and $\hat{y}$ into $\hat{y} = \frac{1}{25}x + b$ and solve for b.

$$\hat{y} = \frac{1}{25}x + b$$

$$6 = \frac{1}{25}(150) + b$$

$$6 = 6 + b$$

$$0 = b$$

So, the linear model that fits the data is $\hat{y} = \frac{1}{25}x + 0$.

Answer $\hat{y} = \frac{1}{25}x$

My Turn!

Find a linear model for the following data for the number of hours spent studying and the final exam score.

Number of Hours Spent Studying	Final Exam Score
4	40
6	50
12	80
18	100

Answers to My Turn!

1. 15
2. 4
3. (0, 2)
4. (0, ½)
5. Answers will vary. (0,3), (1,1), $(2,-1)$
6. 4 pounds
7. 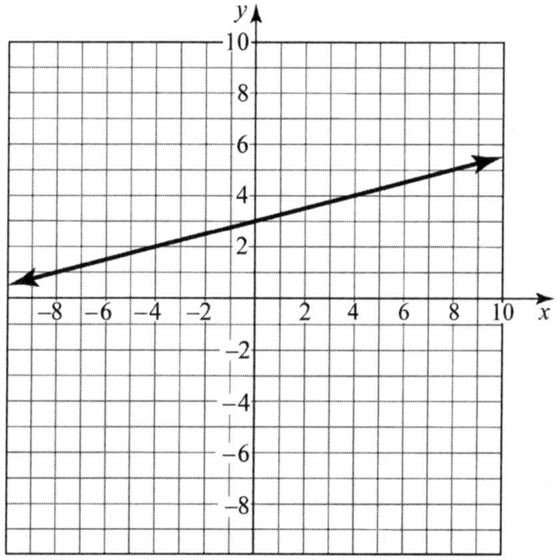

8. 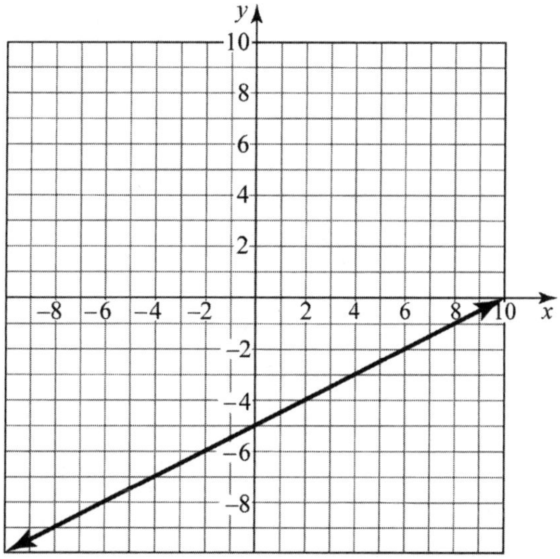

9. $\hat{y} = 5x + 20$

Practice Problems

1. Find the slope of the line that contains the points (3.5, 2.4) and (4.5, 1.6).
2. Find the slope of the line in the following graph.

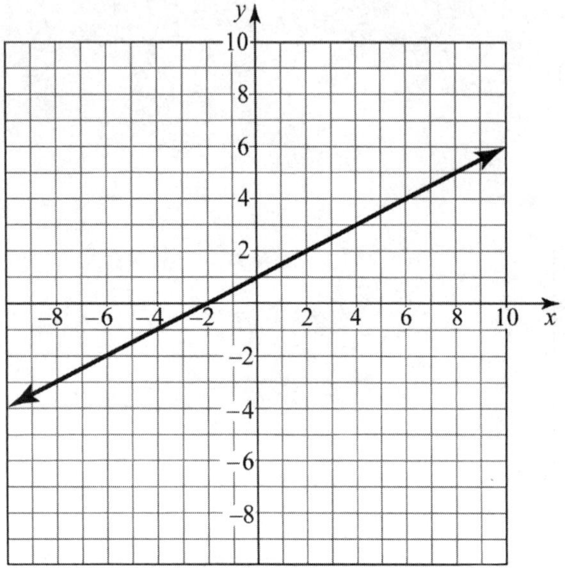

3. Find the y-intercept of the graph shown for Practice Problem 2.
4. Find the y-intercept of the line $\hat{y} = -0.28x + 1.5$.
5. Find three pairs of data values that would be on the line $\hat{y} = -x + 0.2$.
6. The horizontal axis in the following graph represents the number of days that a car is rented, and the vertical axis represents the cost in dollars. How much does it cost to rent a car for 2 days?

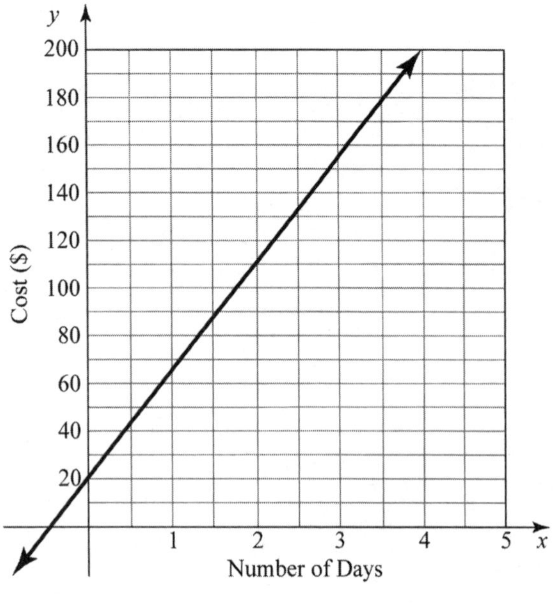

7. Graph $y = -x + 1$ using the slope and y-intercept.

8. Graph $y = \frac{4}{5}x - 3$ by plotting points.

9. Find a linear model for the following data for the number of courses that a student is enrolled in and the number of hours spent watching TV on a Wednesday night.

Number of Courses	Number of Hours Watching TV
2	4
3	3
4	2
5	1

Appendix A: Statistics Symbols

Symbol	Read as
α	Alpha ('ælfə)
β	Beta ('bi:tə)
μ	Mu ('mju:)
$\overline{x}$	*x*-bar
σ	Sigma ('sɪgmə)
ρ	Rho ('rəʊ)
χ	Chi ('kaɪ)
$\hat{p}$	*p*-hat
$\hat{q}$	*q*-hat
$\hat{y}$	*y*-hat
$\overline{d}$	*d*-bar
μ_d	Mu-sub-*d*
s_d	*s*-sub-*d*
$\overline{A}$	*A*-bar, the complement of event *A* (used for probability)
$\approx$	Is approximately equal to
$<$	Is less than
$>$	Is greater than
$\geq$	Is greater than or equal to
$\leq$	Is less than or equal to
df	Degrees of freedom

Appendix B: Units of Measurement

Metric		
Unit of Measurement	**Abbreviation**	**Measures**
Gram	g	Mass (weight)
Meter	m	Length
Liter	l	Volume
Degree Celsius	° C	Temperature

English		
Unit of Measurement	**Abbreviation**	**Measures**
Ounce	oz.	Mass (weight)
Pound	lb.	Mass (weight)
Ton	tn.	Mass (weight)
Inch	in.	Length
Foot (plural feet)	ft.	Length
Yard	yd.	Length
Mile	mi.	Length
Pint	pt.	Volume
Quart	qt.	Volume
Gallon	gal.	Volume
Degree Fahrenheit	° F	Temperature

Appendix C: Conversions between Units of Measurement

Within English System	
16 ounces	1 pound
2240 pounds	1 ton
12 inches	1 foot
3 feet	1 yard
5280 feet	1 mile
2 pints	1 quart
4 quarts	1 gallon

Within Metric System	
Kilo-	1000 basic units (meters, liters, or grams)
Hecto-	100 basic units (meters, liters, or grams)
Deka- (or deca-)	10 basic units (meters, liters, or grams)
Deci-	0.1 basic units (meters, liters, or grams)
Centi-	0.01 basic units (meters, liters, or grams)
Milli-	0.001 basic units (meters, liters, or grams)

Between English and Metric Systems	
1 pound	0.454 kilograms
1 inch	2.54 centimeters
1 mile	1.61 kilometers
1 pint	0.473 liters
1 fluid ounce	0.0296 liters

The conversion from degrees Fahrenheit (F) to degrees Celsius (C) requires the use of the following formula:

$$C = \frac{5}{9}(F - 32)$$

The conversion from degrees Celsius (C) to degrees Fahrenheit (F) requires the use of the following formula:

$$F = \frac{9}{5}C + 32$$

Answers to Practice Problems

Integrated Review 1

1. $d = 12$
2. In the United States, we have presidential elections every four years. So, the possible values for t are 0 (election year), 1, 2, 3, or 4.
3. 16.4 feet
4. 35,000 meters
5. Integers, real numbers
6. 729
7. 7.06×10^{-4}
8. 2.7×10^{5}
9. 97,100,000
10. a) 7 significant digits
    b) 3 significant digits
    c) 1 significant digit
11. 8.67×10^{-19}

Integrated Review 2

1. 24,967.1 kilograms
2. 1.05 pounds
3. $\frac{3}{7}$
4. 0.025
5. 0.7%
6. 8%
7. $\frac{9}{500}$
8. 5%
9. 224
10. 66 subjects

11.

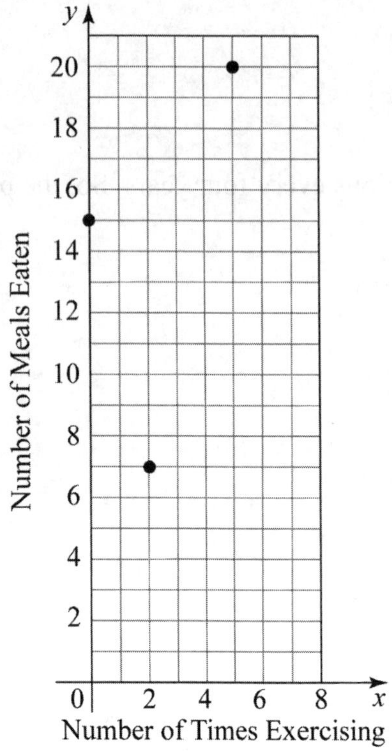

12. 14%
13. 22%

Integrated Review 3

1. 5
2. 14
3. $\frac{16}{85} \approx 0.19$
4. 16
5. 10 and 11
6. 15.97
7. 1.77
8. $\sigma \approx 3.4$
9. 6.5
10. 36.7

Integrated Review 4

1. $\frac{1}{3}$
2. $\frac{79}{70}$

3. $\frac{2}{15}$
4. $\frac{1}{10}$
5. $\frac{1}{24}$
6. $\frac{64}{81}$
7. $\frac{1}{64}$
8. 5.876
9. 3.246
10. 2.625
11. 14.1
12. 362,880
13. 3024
14. 126
15. $A \cap B = \{\text{red}\}$, $A \cup B = \{\text{red, blue, yellow, pink, purple}\}$ and
    $\overline{A} = \{\text{orange, green, purple, pink, white, black}\}$

Integrated Review 5

1. No
2. 1.35
3. 3.85
4. 258.3
5. 967.5
6. 7.1
7. 0.00298
8. 0.3
9. 0.175

Integrated Review 6

1. 0.3 in.^2
2. 10, 11
3. $l > 18$, $v < 60$
4. $z \approx -1.04$
5. $x \approx 98.2$
6. $x \approx 106.6$
7. $z = 4$

Integrated Review 7

1. 0.65
2. 10.105
3. 2.75
4. (3.5,11.9)
5. $0.63 < \mu < 0.92$
6. 25.2 ± 1.1
7. $E \approx 0.057$
8. $E \approx 0.24$
9. $n \approx 140$

Integrated Review 8

1. $z \approx -7.01$
2. $t \approx 3.46$
3. $\chi^2 \approx 161.8$

Integrated Review 9

1. $\overline{p} \approx 0.5533$
2. $z \approx 3.00$
3. $E \approx 0.0959$
4. $t \approx -13.6$
5. $E \approx 0.2$
6. $t = 60$

Integrated Review 10

1. –0.8
2. $\frac{1}{2}$
3. (0, 1)
4. (0, 1.5)
5. Answers will vary. (0, 0.2), $(1, -0.8)$, $(2, -1.8)$
6. $110

7.

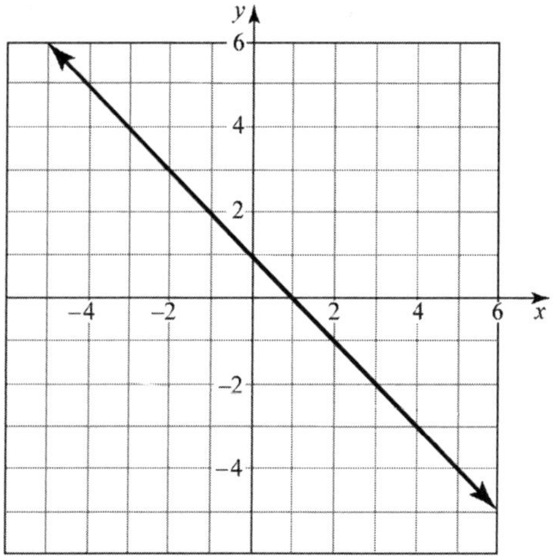

8.

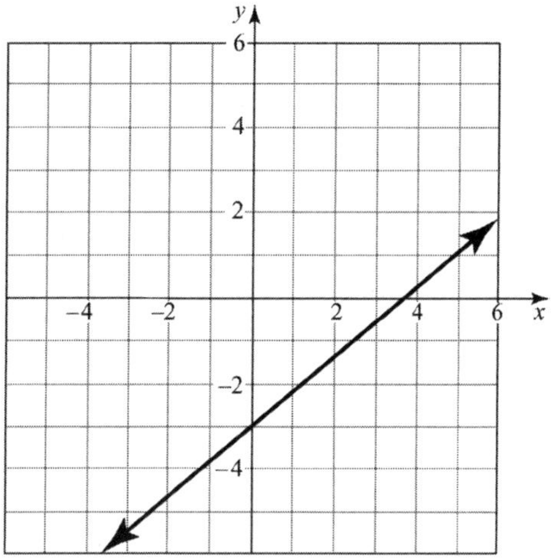

9. $\hat{y} = -x + 6$